Mom! I Learn Addition Using Math-Chess-Puzzles Connection

棋谜式加法

Frank Ho　Amanda Ho

何数棋谜 培训

Ho Math Chess Learning Centre

Table of Contents

How to use this workbook?

- It is important that teachers, tutors, or parents read this "how to" before teaching. Follow the following 3 steps to making tutoring or teaching successfully.

  Step 1: Buy this workbook which costs only equivalently to a one-hour tutoring fee.
  Step 2: Thoroughly read the information in this workbook before Part 1 before tutoring or teaching.
  Step 3: Read the following information and follow them too. If you have any comments or problems while tutoring, please contact us through fho1928@gmail.com.

- The teaching order of addition strategies is important. The teaching order is as follows:
  1. Teach adding by 1, 2, or 3 first.
  2. Teach doubling strategy (2 times table) because the speed of doubling is faster than adding 10.
  3. Teach adding to 10 for 2 of 1 digit. This technique can also be used in subtraction.
  4. Teach 9 + 1 digit such as 2, 3, 4, 5, 6, 7, 8.
      9 + 2 is to give 1 to 9 or split 2 to be 1 + 1, then the answer is 11.

$$9 + 2$$
$$1\ 1 \Leftarrow$$
2 is 1 + 1, the answer is 11.

- Must thoroughly understand the strategies used in this workbook, this workbook does not stress rote teaching or use fingers to count forward or backwards, instead, I emphasize to teach students to use some basic strategies or number senses to get answers.
- Students are required to remember some very basic addition facts such as doubling (double 2 = 2 + 2 = 4, double 3 = 3 + 3 = 6, double 4 = 4 + 4 = 8, double 5 = 5 + 5 = 10, double 6 = 6 + 6 = 12, double 7 = 7 + 7 = 14, double 8 = 8 + 8 = 16, double 9 = 9 + 9 = 18), adding to 10 by 2 of 1-digit(1 + 9 = 10, 9 + 1 = 10, 2 + 8 = 10, 8 + 2 = 10, 3 + 7 = 10, 7 + 3 = 10, 4 + 6 = 10, 6 + 4 = 10, 5 + 5 = 10). It is important to conduct oral tests after student finished each strategy. Do not test each problem in sequential order, but in random order. So, do not ask students what is 1 + 9, 2 + 8, 3 + 7 etc. in sequentially increasing or decreasing order. Ask any problem in random order such as 4 + 6 = ?, 8 + 2 = ?. I have parents told me that they tested their children just last week but when I tested their children, they were not fluent. One of the reasons is the parent gave problem in sequential order.
- Do not let the students drag the study of strategy too long, normally students should show improving results after 8 – 10 hours of study in class and 4 – 6 hours of work at home. If parents do not encourage students to do homework (5 or more pages each week) at home as assigned by the tutor, then the results will be difficult to get.
- Few students only need to do the Part 1, then they could grasp all strategies of addition and move on to more advanced additions such as Part 3 – d + dd etc. Many students need to do Part 2 to reinforce the concepts and do more practices to have a good grasp of the strategies.

- Emphasize the idea of using "reasoning" to get answers after the student understands the strategies such as 6 + 7 = 13 or 7 + 6 = 13 by using 6 + 6 = 12. This is the reason that we have many doubling related problems to teach students how to use the doubling strategies to get other answers. The "reasoning" strategies equally apply to add to 10.

- No chess knowledge is required to work on worksheets, students just follow the dark error of a chess symbol to get numbers or values. The teacher or parent can skip all chess and math integrated materials although it is not recommended. The math, chess, and puzzles integrated worksheets will get students to work more on the brain than the traditional worksheets.

- Not all students in the same grade are at the same math ability level, so do not teach this workbook page after each page. Not all pages need to be finished. I feel very sad if a tutor does not bother to study the teaching ideas of this workbook but just "blindly" ask students to work on these worksheets page after page. Do not go on to a new strategy until the student has mastered each old strategy. If the student still can not master the strategy after finishing all worksheets of this workbook, then most likely, according to my experience, the student and parents must review the student's study ability, methods, and desire.

- What are the reasons that students may not master the addition skills after learning these strategies?.There is a small percentage of students who do not seem to be able to master the addition skills after finishing the Part 1 or Part 2 of its workbook. I can analyze the problems from the three parties which are student, tutor, or guardian.
  (1) The student did not follow the tutor's instructions when working on worksheets. For example, the student is supposed to try to "memorize" the adding to 10 and doubling when working on problems, that was one reason I created many questions so that students will not feel bored easily, but some students did not bother to "memorize" or naturally "instil" into their brain, but just copied answers from the previous answer.
  (2) The student still tried to use fingers especially when the student put hands under the table and gave answers slowly or tried to repeat a tutor's question to buy some time.
  (3) The student has no desire to study arithmetic.
  (4) The student seems to have a very short memory and could understand while being taught, but one or two weeks later, the student would act as almost never being taught at all.
  (5) The tutor continues to use manipulatives other than fingers to teach instead of using reasoning such as 7 + 6 can be logically be answered by using doubling 6 + 6 = 12. Take what you learned and use it on something else is what I want to stress to do additions by using these strategies.
  (6) The tutor used symbols such as drawing circles to do additions, instead of training a student's brain to think logically and abstractly. 4 + 2 can be taught by telling a student to go up 2 floors from the 4th floor.
  (7) The guardian just told the child to do work at home and hardly paid attention o how the work was completed by the student.

- What we like to achieve after using this workbook is to get some students' feedback, hopefully, as follows:

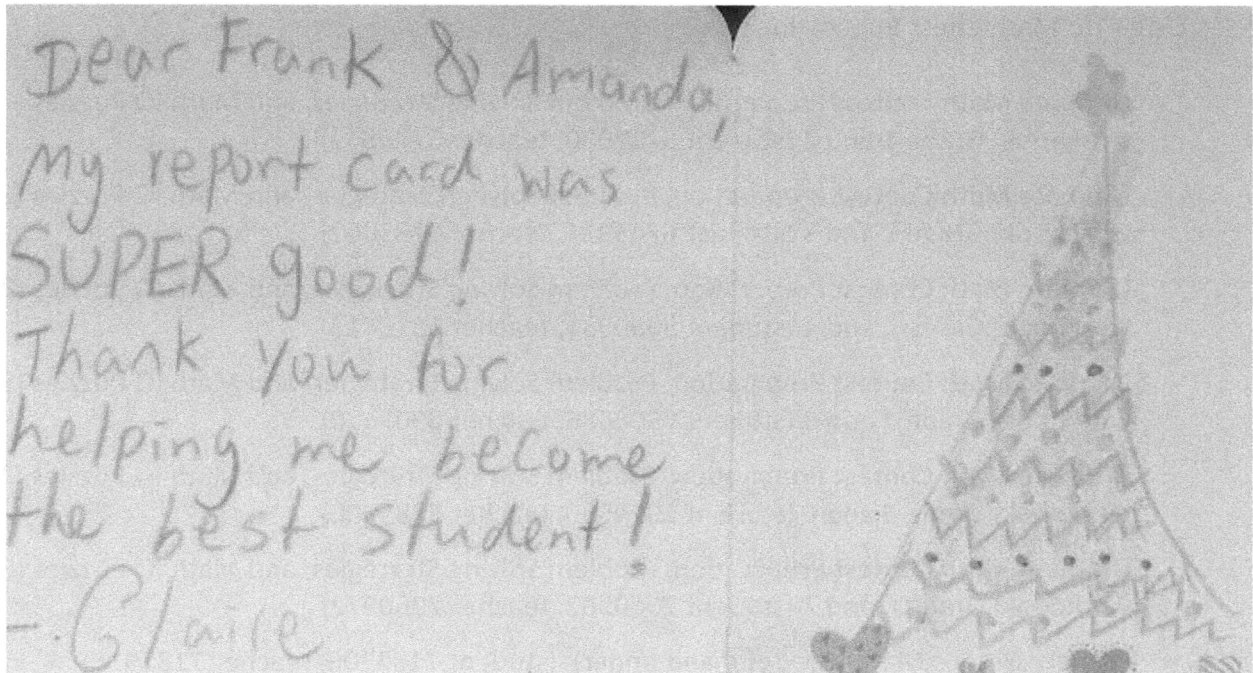

Dear Frank & Amanda,
My report card was SUPER good!
Thank you for helping me become the best student!
- Claire

Ho Math Chess 何数棋谜　妈！我会棋谜式加法啦！
Mom! I Learn Addition Using Math-Chess-Puzzles Connection
Contents include both traditional and Math-Chess-Puzzles combined methods. Extra strength
©2008 – 2018 Frank Ho, Amanda Ho　All rights reserved. www.homathchess.com
Student Name _____ Date _____

- I also have written many other workbooks which are available on www.amazon.com, so there is a continuity of finishing the elementary mathematics by using these workbooks.

Other Ho Math Chess Publications

Ultimate Math Contest Preparation, Problem Solving Strategies, and Math IQ Puzzles (3 in 1 workbook) Grade 1 and 2 (student 6717980, teacher 6720849)

Ultimate Math Contest Preparation, Problem Solving Strategies, and Math IQ Puzzles (3 in 1 workbook) Grade 2 and 3 (student 6745934, teacher 6783062)

Ultimate Math Contest Preparation, Problem Solving Strategies, and Math IQ Puzzles (3 in 1 workbook) Grade 3 and 4 (student 8990731, teacher 6813413)

Ultimate Math Contest Preparation, Problem Solving Strategies, and Math IQ Puzzles (3 in 1 workbook) Grade 4 and 5 (student 6906686, teacher 6907410)

Ultimate Math Contest Preparation, Problem Solving Strategies, and Math IQ Puzzles (3 in 1 workbook) Grade 5 and 6 (student 6979530, teacher 6985906)

Ultimate Math Contest Preparation, Problem Solving Strategies, and Math IQ Puzzles (3 in 1 workbook) Grade 6 and 7 (student 7060502, teacher 7060972)

Primary Grades Math (Grades 4 and under) (student 7182306, teacher 7182916)

Elementary Grades Math (Grades 5 and Up)

Math Chess and Puzzles for Primary Grades (student 9195575, teacher 9195575Junior Kindergarten Math (for Pre-K and junior kindergarten) (student 5738681, teacher 6432094)

Learning Chess to Improve math

Frankho ChessDoku 3 by 3 (student 4233466, teacher 4276722)

Frankho ChessDoku 4 by 4

Frankho ChessDoku 5 by 5

Mom! I Learn Addition Using Math-Chess-Puzzles Connection (student 5919284, teacher 5669603)

Mom! I Learn Subtraction Using Math-Chess-Puzzles Connection (student 658604, teacher 568614)

Mom! I Learn Multiplication Using Math-Chess-Puzzles Connection student 5891187, teacher 5891177)

Mom! I Learn Division Using Math-Chess-Puzzles Connection (student 698270, teacher 5697799)

2018 Christmas

Frank Ho

2017 Preface 前言

Is it insane to have over 500 pages of an addition workbook? The answer is simple, it is not. For a minority of students, they can master the addition strategies taught in this workbook in a short time. Most of the students need to finish a certain quantity to reach fluency. I found out, some students with dyscalculia may need to do multiple times of math worksheets which are not needed for an average student. This is one reason this workbook gets larger than the first version. It is designed for students with different abilities. The other reason why this workbook is so huge is that this revised Addition workbook integrates the following previously published three workbooks:

Mom! I Learn Addition Using Math-Chess-Puzzles Connection
Learning Calculation without Counting Fingers
Math Entrance Examination for Private School

The new and revised Addition workbook particularly stresses the importance of teaching the Addition strategies in the following order:

Strategy 1: Adding 1, 2, or 3
Strategy 2: Doubling
Strategy 3: Adding 2 numbers to 10
Strategy 4: 9 Plus 1 after the number split into 1 and (number -1)

The reason for integrating all 3 books into one workbook is so that students can achieve many learning purposes without the need to buy 3 workbooks. The other reason is to train students to do calculations without counting fingers.

The following is a preface which I wrote for the workbook *Learning Calculation without Counting Fingers*.

The design of **Ho** Math Chess worksheets encourages students to do calculations without using their fingers to count numbers. Almost none of our A students and students participating in math competitions use their fingers to do basic calculations and their ability of mental math is strong than those who use fingers to calculate. The main reason is, without using fingers; students will develop a strong number sense and are able to see steps ahead when working on math problems.

The following gives a list of potential problems for some elementary students who continue to use fingers to do calculations might encounter when going to higher grades:

- Difficult and slow to find answers which are related to problems requiring work backwards such as subtractions or finding quotients. For example, it takes finger calculating students much longer to find what is 12 - 9? When doing simple divisions such as 24 ÷ 2, 36 ÷ 3 finger

- calculating, students cannot do them mentally without using pencil and paper.

- Have hard time to find product of primes of a number; slow to find an answer and have hard time to see if a number is divisible by a certain number using divisibility rules; difficult to find GCF or LCM and later LCD; difficult and slow to reduce a fraction; have hard time to do the conversions between fractions, decimals, and % etc.

- Unable to see what is the trick to add certain type of questions such as 99 + 7 + 8 + 4 + 6 + 1 + 3 + 2 etc. Students who do not use fingers can scan this problem and they know the answer almost immediately and automatically.

- Because of the slow speed of doing basic calculations and the lack of mental math ability, the student has less chance to get in the enrichment math program due to competitiveness. By the same token, the student has less chance to do reasonably well in any math contests which stress calculations speed and the calculator is not allowed. Students will do poorly when oral math problems are presented in any math contests.

- Students will continue to experience difficulties when going to high schools with those problems requiring mental math ability to quickly recall four basic computation facts. Problems such as finding the square roots; adding like terms; factoring trinomials using cross-multiplication method; 30-60-90 special degree; graphing $2x + 3y = 6$ etc.

When we talk about calculation without counting fingers, we are not saying the only reason is so that students can recall arithmetic facts quickly. From the above observations, we can see that mental math or calculate without counting fingers is not limited to the scope of being a human calculator of doing some calculations in fast speed. To acquire the mental math skills, so it truly benefits students, the students must learn the skill of reversing calculating like how to get factors of a number; how to quickly find prime factors of a number; how to find a number if it has a perfect number as a factor; how not only do calculation in linear fashion; but also know how to do calculations in cross or vertical way quickly and be able to see patterns. Mental math also helps the student do trinomial factoring when they are in higher grades such as 9 or 10.

When in high school, students with math deficiencies often find it is difficult for them to go back to work on addition, subtraction, subtraction, or multiplication again since it is embarrassing for them to work on grade 3 math while they are already in grade 10.

With the above in mind, one purpose of this workbook is to create an environment where students are encouraged to think more even when they are doing calculations. This workbook encourages students to do calculations without counting their fingers, so the calculations are totally done in their brains without relying on any manipulative. The requirement for students to write the numbers to arrive answers will help them reinforce of memorization of number facts.

These mental calculations worksheets will help students develop strong mental math ability. As consequence, this helps students do well later in their SSAT, SAT and increase the chance of getting into private schools if they wish to. The chance of getting into an enrichment math program in high schools is also higher.

April 2017

Ho Math Chess worksheets and traditional worksheets

Frank Ho, Amanda Ho

Ho Math Chess Tutoring Centre
Vancouver, Canada

What's wrong with the traditional drill computation? From tutoring point of view, there is nothing wrong to give them to children for practice on fluency and grasp of basics. However, there is something wrong from children's point of view that is they are boring, dull and not fun. Why do children feel that way? Well, the time has changed but the format of traditional computation worksheets has not caught up with the pace of the society. The future belongs to a generation who understands how to process information and the information might include digits, bytes, numbers, graphics, images, languages, symbols, equations etc. Children today might chat with others on the internet while downloading or uploading files and viewing movie clips at the same time. Multitasking and the multi-way of processing information seem to have come as second nature to children, but is our computation format reflecting the way that children are living today? Certainly not, this is one reason why children feel so bored and lost interest to continue to work on the same" old" style of computation worksheets.

These simple monotonic basic number facts computation worksheets are no longer reflecting the real world which the young generation is facing today or will be living in the future. The computing world children are facing today is much like a rich tapestry, where diversified fabric and colours are integrated. Children today are absorbing not just numbers but an array of information like image, sound, music, symbols, spatial information, or even abstract ideas all bundled together and delivered through many types of media. Children today are not happy just working on pure number drill without any other stimulus or motivator. After realizing the importance of having fun while learning, Ho Math Chess has been embarked on an important teaching philosophy that is to integrate chess and puzzles into math worksheets so that children can learn math while having fun.

Ho Math Chess 何数棋谜　妈！我会棋谜式加法啦！
Mom! I Learn Addition Using Math-Chess-Puzzles Connection
Contents include both traditional and Math-Chess-Puzzles combined methods. Extra strength
©2008 – 2018 Frank Ho, Amanda Ho　All rights reserved. www.homathchess.com
Student Name _____ Date _____

The values of chess pieces used in this workbook are listed as follows:

	Pawn	Knight	Bishop	Rook	Queen	king
Ho Math Chess	⬇	✛	⤡	⬌	✳	✳
Traditional chess	♟	♞	♝	♜	♛	♚
Value	1	3	3	5	9	0

The following offers a brief comparison between Ho Math Chess worksheets and traditional worksheets

	Traditional worksheets	Ho Math Chess worksheets
Computing Direction	Mostly left-right or top-down	Multi-direction math and chess integrated. Chess queen has 8 directions. The reason for training children to calculate in multi-way is as follows: Math computation is not just one way of computing from left to right or top to down. Many fractions require reducing in a diagonal way such as $\frac{4}{3} \times \frac{9}{8}$.
Concept	Mostly one concept	Multi-concept. For example, many work backwards problems are presented in addition worksheets to train students to get answers by using calculating backwards strategy.
Abstract symbols	Very little abstract symbols are used in arithmetic computations.	Chess abstract symbols are used starting from pre-k.
Subject	Only math	Multi-subject: math, chess, puzzles, and brainpower
Entertainment	No entertainment value	Offering entertainment value through puzzle-like problems. Many times, no problems are present obviously, thus Ho Math and Chess worksheets train student's thinking skills while working on pure computation worksheets.
Computing	Only number crunching is offered	Students need to think a lot more to find out what is the question when working on a simple calculation problem.
Manipulative	Short term value	Chess has life-long value.
idea	Worksheets have no new ideas.	Innovative worksheets train students to have creative ideas.
Contents	Many similar worksheets	World first, unique and the only one.

Ho Math Chess workbook is a multi-function workbook, it trains children not only their basic computing ability but also trains them to be an astute data warehouse manager or an excellent data miner by developing their problem-solving ability and critical thinking skills.

Ho Math Chess workbook provides education and entertainment value to get young children involved in the future world they will be facing.

Ho Math Chess worksheet layout simulates
a computer screen and a cell phone screen.

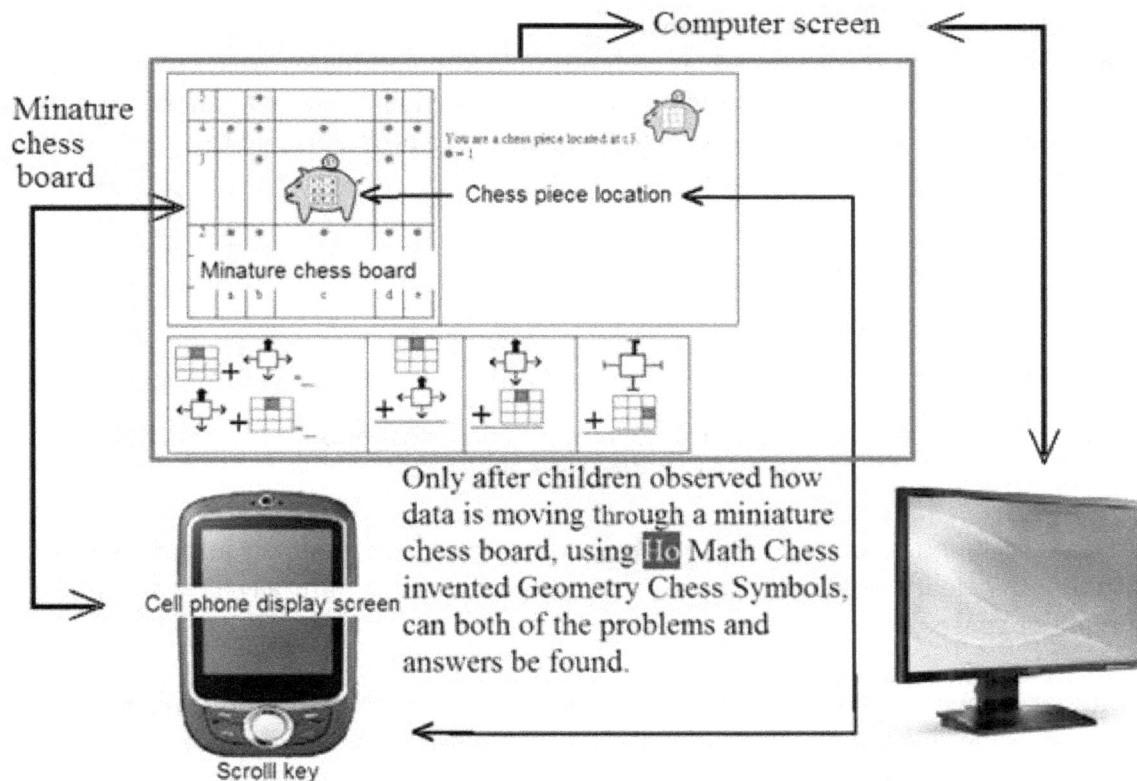

Minature chess board

Computer screen

You are a chess piece located at c3.
● = 1

Chess piece location

Minature chess board

a b c d e

Cell phone display screen

Scrolll key

Only after children observed how data is moving through a miniature chess board, using Ho Math Chess invented Geometry Chess Symbols, can both of the problems and answers be found.

Benefits of using Ho Math Chess workbook

We have found that the traditional worksheets seem to present difficulties for young children to learn math especially 4-year-old, many of them end up using fingers as a mechanism to count. This workbook is created very differently from the traditional workbook. It will eliminate the tendency for children to use fingers and learn the skills of addition with a shorter learning curve. The other most important difference is this workbook is much more fun than working on the traditional worksheets and Ho **Math Chess** requires students to "think" more than traditional worksheets. The other major differences are as follows:

1. For the traditional math workbooks, children only need to write the answers. For example, for additions, children only need to write the answer of additions but no need to write what numbers have been added together such as addends unless the problems are word problems solving type. In this workbook, most of the problems require children to write addends so that children have opportunities to have the effect of "reciting" the problems and thus reinforcing their memory. An example is given as follows:

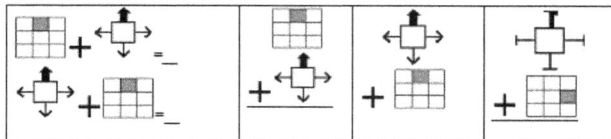

2. Children will feel more fun if working on this workbook since most of our problems having "puzzles" built in.

3. Many additions techniques such adding up to ten, doubling etc. are gradually introduced to children so that they learn addition in an easy and intuitive way and naturally increase the ability of number sense.

4. One reason that many children have math problems is that they cannot do the reverse thinking, connecting and organizing information, or analyzing problems. This workbook is created to train students to work on computation worksheets, yet students will also get the same benefits as they were working on word problems. This is because our worksheets are created such that students will have to

   • Analyze to find out how to solve the problem by following abstract symbols.
   • Visually connect pieces of information together by following symbols.
   • Organize information by table lookup.
   • Put them together by the following the logic.
   • Follow direction (in chess moves).
   • Find ways around to look for information (convergent and divergent).

The process of operating instructions provides fun and challenging for children.

This workbook improves children's math computing and problem-solving abilities and at the same time, increases their brain power.

Frank Ho
Amanda Ho

January 2012

Student Name _____ Date _____

使用何数棋谜教材的好处

加拿大何数棋谜培训中心

創辦人何数棋（Frank Ho）, 何数谜（Amanda Ho）

www.mathandchess.com

只见棋谜不见题　　劝君迷路不哭涕
数学象棋加谜题　　健脑思维真神奇

今天儿童面对的世界是学习如何处理数字,图形,资料搜寻,音影上下载,资讯比较,分类等资讯.这些活动实际已成為儿童生活的一部份.所以如果说学数学就是计算数字就错了.学数学的另一个目的就是学习如何利用数字资讯去解决问题及培养创造力.但是**传统式数学的计算练习题却完全没跟上科研已经改变了儿童面对的世界**.

何数棋谜

棋谜式健脑思维趣味数学结合高科技操作图解

只见棋谜不见题　劝君迷路不哭涕
数学象棋加谜题　健脑思维真神奇

Ho Math and Chess worksheet layout simulates
a computer screen and a cell phone screen.

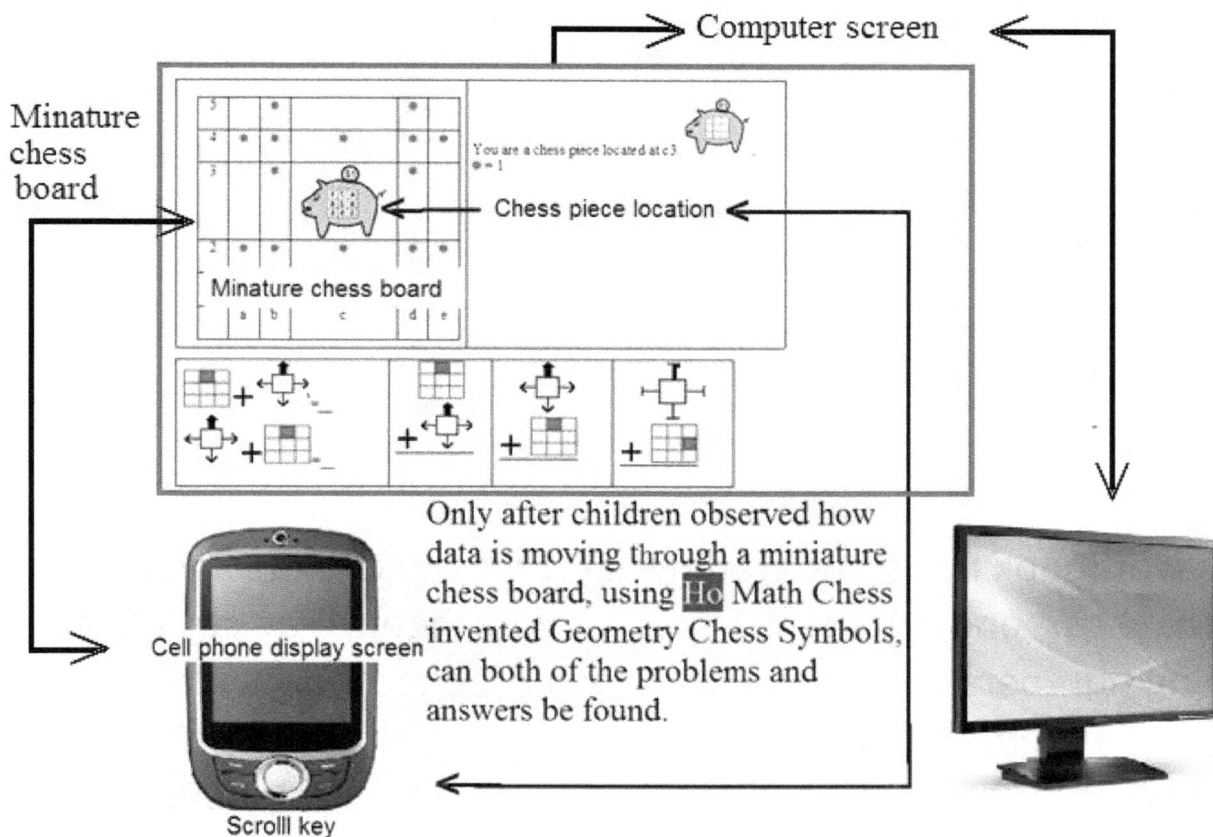

Computer screen

Minature chess board

You are a chess piece located at c 3

Chess piece location

Minature chess board

Cell phone display screen

Scrolll key

Only after children observed how
data is moving through a miniature
chess board, using Ho Math Chess
invented Geometry Chess Symbols,
can both of the problems and
answers be found.

a

Chess symbols and their mathematical values

Symbols of chess pieces	Names of chess pieces	Mathematical values
✻ ✻ ♛ ♛	Queen	9
✢ ✢ ♜ ♜	Rook	5
✕ ✕ ♝ ♝	Bishop	3
✢ ✢ ♞ ♞	Knight	3
♈ ♈ ♙ ♟	Pawn	1
✳ ✳ ♔ ♚	King	0

This workbook has some traditional style worksheets and our unique math, chess, and puzzles integrated worksheets. The strategies of adding numbers are offered and each problem's difficulty increases in small increment and in a very organized pattern to get students familiar with the relationship between numbers and develop their number sense.

Prerequisites before working on additions

Before a student could work on "real" addition, there is some arithmetic knowledge that a student must possess such as the following:

- To be able to recite from 0 to 100 forward and also backwards fluently.
- To be able to write digits from 0 to 9 in correct writing strokes.
- To know the rank of numbers such as 3 is larger than 2 by 1 and 4 is one less than 5 etc.
- To know the names of the shapes.

If a student has some difficulties of achieving the level of the above arithmetic knowledge, we have two workbooks *Junior Kindergarten Math* and *Math Chess and Puzzles for Primary Grades* for students for them to work on.

The workbook *Math Chess and Puzzles for Primary Grades* can also be used along with this workbook, so the student will not feel too bored when just working on computations all the time.

Counting numbers and whole numbers

Numbers starting with 1 and then counting forward by adding 1 such as 1, 2, 3, 4, 5, 6, 7, … and so on are called counting numbers (natural numbers). Counting numbers are also called whole numbers. The difference between counting numbers and whole numbers is that the whole numbers include 0 but the counting numbers do not include 0. In this book, whenever the word "number" is mentioned, it means the whole number.

Fill in the following □ by a number.

1, 3, 5, 7, □, 11, 13, □

Replace the following "?" by a counting number.

The numbers starting at 2 and added by 2 each time are called _____ numbers. 0 can be considered as an even number for a computational purpose like checking if a number is divisible by 2.

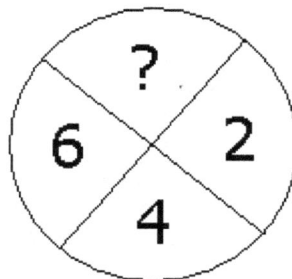

Replace the following ? with a counting number.
The numbers starting at 1 and added by 2 each time are called _____ numbers.

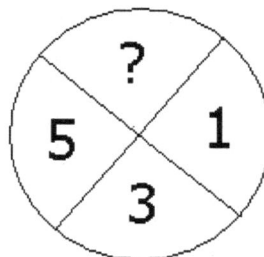

Writing d + d number sentence (Commutative property) 交换率

Fill in each ☐ with a number.

Math and Chess sentence	Number sentence.	Comments
♟♟♟♟♟ + ♟♟	$5 + 2 = 7$	**5 plus 2 is 7.** 5 and 2 are (equals) 7. The sum of 5 and 2 is (equals) 7. 5 added to 2 is (equals) 7. The sum of 5 and 2 is equal to 7. Addend + Addend = Sum **Think to add from left to right (Think 5 then add 2).**
♟♟ + ♟♟♟♟♟		Think to add 2 to 5 (from right to left).
♟♟♟♟♟♟♟ + ♟♟		Think to add from left to right (Think 7 then add 2).
♟♟ + ♟♟♟♟♟♟♟		Think to add from right to left (Think 7 then add 2).
♟♟♟♟♟♟♟♟♟ + ♟♟		Think to add from left to right (Think 9 then add 2).
♟♟ + ♟♟♟♟♟♟♟♟♟		Think to add from right to left (Think 9 then add 2).
♟♟♟♟♟♟♟ + ♟♟♟		
♟♟♟♟ + ♟♟♟♟♟♟♟		
♟♟♟♟ + ♟♟♟♟♟♟♟♟		
♟♟♟♟♟♟♟♟♟ + ♟♟♟		

Commutative law (a + b = b + a) of adding up to 5 交换律
The teacher can use a pile of pencils as a total to split the total.

Fill the following ☐ with a number.

0	+	☐	=	♖	=	♖	+	☐
♙	+	☐	=	♖	=	4	+	☐
2	+	☐	=	♖	=	♘	+	☐
♘	+	☐	=	♖	=	2	+	☐
4	+	☐	=	♖	=	1	+	☐
♖	+	☐	=	♖	=	♔	+	☐

Student Name _____ Date _____

Strategies for d + d addition 加法的策略

The following addition strategies reveal how additions could be done using a few strategies without using fingers or boring repetitions. With the above in mind, the teacher or parent must be very careful not to teach children with a wrong strategy, otherwise, children could get confused.

For example, 8 + 7 = double 7 plus 1usind doubling strategy, not to teach children to use 8 + 7 = 8 + 2 + 5 = 10 + 5 = 15. It is alright to teach addition strategy to children, but children still need to master one technique and continue to use the same strategy until they can calculate without any hesitation.
Teach children so they know 8 + 7 = 7 + 8 = double 7 + 1 = 15.

	1	2	3	4	5	6	7	8	9
1	double 1 (Add the number itself.)								
2	+ 1	double 2							
3	+ 1	+ 2	double 3						
4	+ 1	+ 2	+ 3	double 4					
5	+ 1	+ 2	+ 3	1 away from double 5	double 5				
6	+ 1	+ 2	+ 3 (later remember 3 x 3 =9)	Adding to 10	double 5 +1	double 6			
7	+ 1	+ 2	Adding to 10	Adding to 10 +1	double 5 + 2	double 6 + 1	double 7		

Student Name _____ Date _____

	1	2	3	4	5	6	7	8	9
8	+ 1	Adding to 10	Adding 10 + 1	Adding 10 + 2 (later remember 3 of 4 =12)	Double 5 + 3	double 6 + 2	double 7 +1	Double 8	
9	Adding to 10	9 + 1+1 (note that 2 = 1+1)	9 + 1+ 2 (note that 3 = 1+2)	9 + 1+ 3 (note that 4 = 1+3)	9 + 1+ 4 (note that 5= 1+4)	9 + 1+ 5 (note that 6= 1+5)	9 + 1+ 6 (note that 7= 1+6)	9 + 1+ 7 (note that 8 = 1+7)	double 9

English names for numbers (For students who can read and write English.)

Large numbers are for advanced students only. These worksheets depend on a student's English ability, so do not enforce it.

Whole number	English words	Comments
0		
1		
2		
3		
4		
5		
6		
7		
8		
9		
10		
11		
12		

13		
14		
15		
16		
17		
18		
19		

English names for numbers

Whole number	English words	Comments
20		
30		
40		
50		
60		
70		
80		
90		
21		
25		If a number is in the range of 21 to 99 and the second digit is not zero, a hyphen is used to separate the two words.
36		
47		
86		

99		
100		No "s" needs to be added to "hundred".
200		
900		

English names for numbers

2,000		
9,000		
19,000		No "s" needs to be added to "thousand".
20,000		
24,000		
79,000		
100,000		
1,000,000		

Number order and rank

Insert >, <, or =.

Students must know the order and ranking of numbers and be able to recite fluently forward and backwards up to 20 before using counting skills in the brain (not fingers) and addition. Students should work on less than 20 numbers first.

1.	**5**	☐	**3**	2.	4	☐	3
3.	**6**	☐	**7**	4.	**7**	☐	**5**
5.	**5**	☐	**3**	6.	**8**	☐	**6**
7.	**7**	☐	**5**	8.	**2**	☐	**3**
9.	**9**	☐	**8**	10.	**9**	☐	**10**
11.	11	☐	10	12.	16	☐	16
13.	13	☐	17	14.	15	☐	18
15.	8	☐	7	16.	17	☐	14
17.	27	☐	27	18.	13	☐	15
19.	27	☐	29	20.	23	☐	21
21.	33	☐	30	22.	26	☐	28
23.	43	☐	34	24.	31	☐	29
25.	57	☐	67	26.	34	☐	37
27.	51	☐	69	28.	75	☐	77
29.	39	☐	41	30.	32	☐	36
31.	34	☐	86	32.	67	☐	76
33.	58	☐	54	34.	38	☐	83
35.	80	☐	82	36.	32	☐	27

Student Name _____ Date _____

Order the following numbers from smallest to largest.

5	8	3	6	_____
6	2	7	1	_____
4	9	5	3	_____
3	5	1	7	_____
11	15	13	18	_____
34	44	6	54	_____
51	61	31	21	_____
36	42	65	41	_____
46	24	26	43	_____
25	34	64	52	_____
57	53	50	56	_____
35	38	32	37	_____
42	22	52	32	_____
21	41	51	31	_____
42	56	53	27	_____
50	64	36	19	_____
57	36	71	50	_____
47	38	92	65	_____
36	64	38	60	_____

Order the following numbers from greatest to smallest.

5	8	3	7	_____
2	5	6	3	_____
7	9	4	5	_____
2	1	7	6	_____
3	6	1	8	_____
11	16	13	15	_____
47	49	41	46	_____
32	62	72	42	_____
23	93	73	53	_____
45	84	85	63	_____
52	54	74	37	_____
45	75	63	28	_____
47	37	34	86	_____
26	53	75	70	_____
19	76	85	46	_____
45	86	36	58	_____
73	27	48	43	_____
78	46	27	58	_____
57	38	96	34	_____

Part 1: Adding 123, doubling, adding ten, and 9 + d strategies

Most educators would think that learning computation is just "drill to kill' but the truth is because most workbooks do not provide any "thought-provoking" type of questions. The computation type of workbooks can be written in a way to train students' problem-solving skills, much like puzzles can be used to train a student's problem-solving skills. Ho Math Chess computation workbooks are written to increase a student's computation ability, but also to increase a student's problem-solving ability.

Students need to thoroughly understand the strategy of "123DT" that is to **add 1, 2, or 3** to a number, **doubling (D)**, **adding ten (T)** and be able to recall results like recalling students' own phone numbers. Once they master these three basic skills then they will have little problems to add multi-digit numbers. This type of problems also trains the student's ability to work backwards.

Many children are not able to make progress or fall behind when using the "regular" math materials which are used at day schools or materials bought from stores. We have produced this special student math materials for those who need extra help.

We have a workbook called Frankho ChessDoku 3 by 3 which integrates math, chess, and Sudoku all in one workbook, Students love to work on this workbook to train their logical thinking skills and also have fun. This workbook can be used to complement the training of training a student's 1, 2, 3 addition skills. This workbook can be purchased from www.amazon.com.

The remaining part of this section contains a 7-day of practice worksheets of adding 1, 2, 3, doubling, adding to ten and 9 plus strategies. Some students could become proficient in additions, but the majority of students need more practices.

Mom! I Learn Addition Using Math-Chess-Puzzles Connection

Contents include both traditional and Math-Chess-Puzzles combined methods. Extra strength

Student Name _____ Date _____

Mixed counting 1, 2, 3 forward d + d (Start with a bigger number and count on dots).

(♔ = 0, ♙ = 1, ♗ = 3, ♞ = 3, ♖ = 5, ♛ = 9)

5 + . ♙	. ♙ + 5	6 + .1	7 + ..2	8 +. ♙
.1 + 8	7 + ...♞	6 + ...3	7 + .. 2	5 + ..2
6 + ..2	...3 + 6	.1 + 7	7 + ..2	.1 + 8
8 + .1	5 + ..2	6 + ..2	..2 + 7	7 + . ♙
. ♙ + 8	7 + ..2	6 + ..2	. ♙ + 6	8 + ..2

Mom! I Learn Addition Using Math-Chess-Puzzles Connection

Contents include both traditional and Math-Chess-Puzzles combined methods. Extra strength

Student Name _____ Date _____

Mixed counting 1, 2, 3 forward d + d (Start with bigger number and count on dots).
(♔ = 0, ♙ = 1, ♗ = 3, ♘ = 3, ♖ = 5, ♕ = 9)

6 + .1	.1 + 7	6 + .1	7 + ..2	8 + .♙
.♙ + 8	7 + ..2	6 + ...♗	7 + .. 2	6 + ...♘
6 + ..2	...♘ + 6	...3 + 7	7 + ..2	..2 + 8
7 + ..2	6 + ..2	6 + ...3	...♗ + 7	7 + .1
..2 + 8	2 + ..7	7 + ..2	...♘ + 6	8 + .♙

Mixed counting 1, 2, 3 forward and backward of d + d or d − d
(♔ = 0, ♙ = 1,　♗ = 3, ♘ = 3, ♖ = 5, ♕ = 9)

5	..2	6	7	8
+ . ♙	+ 5	- .1	+ ..2	+ ..2
.1	7	6	9	8
+ 8	+ ..2	+ ...2	- ..2	- ..2
6	...3	.1	7	. ♙
+ ..2	+ 6	+ 9	- ..2	+ 8
8	5	6	..2	7
+ . ♙	- ..2	+ ..2	+ 7	+ .2
.1	8	6	. ♙	8
+ 8	- ..2	+ ..2	+ 6	- .1

Student Name _____ Date _____

Mixed counting 1, 2, and 3 forward and backwards of d + d or d − d
(♔ = 0, ♙ = 1, ♗ = 3, ♘ = 3, ♖ = 5, ♕ = 9)

6		2	♕	7	8
+ 1		+ 5	- 2	+ 2	+ 1
♙		6	6	7	♕
+ 8		+ 2	+ 3	- 2	- 2
6		3	♙	7	1
+ 2		+ 6	+ 7	- ♙	+ 8
8		♖	6	2	6
+ ♙		- 2	+ 2	+ 7	+ 1
♙		8	6	1	7
+ 8		- 2	+ 2	+ 6	- ♙

Mom! I Learn Addition Using Math-Chess-Puzzles Connection

Contents include both traditional and Math-Chess-Puzzles combined methods. Extra strength

Student Name _____ Date _____

Mixed counting 1, 2, and 3 forward and backwards of d + d or d - d
(♔ = 0, ♙ = 1, ♗ = 3, ♘ = 3, ♖ = 5, ♕ = 9)

3	♗	8	7	8
+ 1	+ 5	- 2	+ ♙	+ 1
2	6	♗	8	♕
+ 7	+ ♙	+ ♗	- 2	- 1
6	2	2	7	1
+ ♗	+ 6	+ 7	- 2	+ 8
8	5	6	♙	6
+ ♙	- 3	+ ♗	+ 8	+ 3
♙	8	6	♗	7
+ 7	- ♗	+ 2	+ 6	- 2

Student Name _____ Date _____

Mixed counting 1, 2, 3 forward and backwards of d + d or d - d
(♔ = 0, ♙ = 1, ♗ = 3, ♘ = 3, ♖ = 5, ♕ = 9)

6	3	9	7	8
+ ♗	+ 5	- 2	+ 2	+ 1
♙	♗	6	7	9
+ 8	+ 6	+ 3	- 2	- 2
6	♞	♗	7	♙
+ 2	+ 6	+ 5	- 2	+ 8
4	6	5	2	6
+ 3	- ♗	+ 3	+ 7	+ ♞
♞	8	6	3	7
+ 5	- 3	+ 2	+ 6	- ♗

Mixed counting 1, 2, 3 forward and backwards of d + d or d - d
(♔ = 0, ♙ = 1, ♗ = 3, ♘ = 3, ♖ = 5, ♕ = 9)

8 + ♙	3 + ♖	8 − 2	6 + 2	7 + 1
1 + 6	7 + 2	♖ + 3	♕ − 2	8 − 2
7 + 2	3 + 6	♙ + 8	6 − ♙	♙ + 7
♖ + 4	7 − 2	♖ + 2	2 + 4	6 + 1
♙ + 8	♕ − 2	6 + 2	2 + 7	8 − ♙

Mixed counting 1, 2, 3 forward and backwards of d + d or d - d
(♔ = 0, ♙ = 1, ♗ = 3, ♘ = 3, ♖ = 5, ♕ = 9)

7 + 2	2 + 7	8 + ♙	1 + 8	6 + ♘
3 + 6	♖ + ♘	♘ + 5	♖ + 4	4 + 5
5 + 2	2 + ♖	♙ + 5	♖ + 1	1 + 4
4 + ♙	4 + 2	4 + 2	♙ + 4	4 + 3
♘ + 3	3 + 2	2 + ♘	1 + 3	♘ + 1

Ho Math Chess 何数棋谜　妈！我会棋谜式加法啦！

Mom! I Learn Addition Using Math-Chess-Puzzles Connection

Contents include both traditional and Math-Chess-Puzzles combined methods. Extra strength

Student Name _____ Date _____

Mixed counting 1, 2, 3 forward and backwards of d + d or d - d
(♔ = 0, ♙ = 1, ♗ = 3, ♘ = 3, ♖ = 5, ♕ = 9)

8	4	8	6	7
+ 1	+ 5	- ♗	+ 3	+ ♙
♗	7	♖	9	8
+ 6	+ 2	+ ♗	- 3	- ♗
6	♗	♙	6	2
+ ♗	+ 6	+ 8	- ♙	+ 7
7	7	♖	♗	6
+ 1	- 2	+ 2	+ 4	+ 3
♙	9	6	♙	8
+ 8	- ♗	+ 2	+ 6	- 2

Mom! I Learn Addition Using Math-Chess-Puzzles Connection

Contents include both traditional and Math-Chess-Puzzles combined methods. Extra strength

Student Name _____ Date _____

Counting forward by 2's (even numbers or odd numbers)

2	4							
13	15							
54	56							
78	80							
33								
78								
62								
19								
27								
		84						
26	28							
45	47		51					
53	55				63			
14	16	18						
24	26		30					
63	65						77	

Counting backwards by 2's (even numbers or odd numbers)

20	18	☐	☐	☐	☐	☐	☐	☐
23	21	☐	☐	☐	☐	☐	☐	☐
61	59	☐	☐	☐	☐	☐	☐	☐
44	42	☐	☐	☐	☐	☐	☐	☐
32	30	☐	☐	☐	☐	☐	☐	☐
56	54	☐	☐	☐	☐	☐	☐	☐
77	75	☐	☐	☐	☐	☐	☐	☐
99	97	☐	☐	☐	☐	☐	☐	☐
78	76	☐	☐	☐	☐	☐	☐	☐
16	14	12	☐	8	☐	☐	☐	☐
29	27	25	☐	☐	☐	☐	☐	☐
54	52	☐	48	☐	☐	42	☐	☐
25	23	21	☐	☐	15	☐	☐	☐
81	79	77	☐	☐	☐	69	☐	☐
48	46	☐	42	☐	38	☐	☐	☐
41	39	☐	☐	33	☐	☐	27	☐

Counting forward by 5's (for working on clock problems)

5	10	15	20	25	30	35	40	45
15	20	25	30	35	40	45	50	55
20	25	30	35	40	45	50	55	60
41	46	51	56	61	66	71	76	81
35	40	45	50	55	60	65	70	75
60	65	70	75	80	85	90	95	100
32	37	42	47	52	57	62	67	72
47	52	57	62	67	72	77	82	87
52	57	62	67	72	77	82	87	92
15	20	25	30	35	40	45	50	55
35	40	45	50	55	60	65	70	75
45	50	55	60	65	70	75	80	85
27	32	37	42	47	52	57	62	67
21	26	31	36	41	46	51	56	61
48	53	58	63	68	73	78	83	88
43	48	53	58	63	68	73	78	83

Counting backwards by 5's (for working on clock problems)

55	50	□□	□	□	□	□	□	□
70	65	□□	□	□	□	□	□	□
46	41	□□	□	□	□	□	□	□
94	89	□□	□	□	□	□	□	□
87	82	□□	□	□	□	□	□	□
40	35	□□	□	□	□	□	□	□
62	57	□□	□	□	□	□	□	□
57	52	□□	□	□	□	□	□	□
76	71	□□	□	□	□	□	□	□

□	□	65	□	55			
□	80	75	□	□			
□	□	□□	57	□	□	42	□	□
□	35	□□	□	15	□			
□	□	91	□	□	□	71		
□	□	□□	76	□	66			
□	□	□□	□	32	□	□	17	□

Addition strategy: reverse calculation of sums 倒算和

Students learn to add to a number by circling the numbers, then learn to partition a number. 先凑後拆
These strategies also help students learn our Frankho ChessDoku 3 by 3 puzzles.

Adding to 3. 凑 3
Add digits of each number then write each sum beside that number.
Circle the following numbers when the sum of their digits is 3.
Think 1 + 2 = 2 + 1 = 3
If the students have difficulties in doing adding 4, then just let them skip the questions such as 5 + 4. The goal is to master adding 1, 2, and 3 first.

21	32	43	54
31	42	53	54
41	52	21	21
21	62	63	
51	72		
61			
21			
71			
81			

12	23	34	45
13	24	35	45
12	21	21	21
14	25	35	
15	26		
12	27		
16	21		
17			
18			

Adding to 4 凑 4

Add digits of each number then write each sum beside that number.
Circle the following numbers when the sum of their digits is 4.

Think 2 + 2 = 4, 1 + 3 = 3 + 1 = 4

If the students have difficulties in doing adding 4, then just let them skip the questions such as 5 + 4. The goal is to master adding 1, 2, and 3 first.

21		32	43	54
31		42	53	31
41		31	31	
51		52	63	
31		62	22	
61		31		
71		72		
31				
81				

12	23	34	22
13	22	13	45
14	24	35	13
15	13	35	
13	25	13	
16	26		
17	13		
13	27		
18	13		

Student Name _____　　Date _____

Adding to 5 凑 5

Add digits of each number then write each sum beside that number.
Circle the following numbers when the sum of their digits is 5.

Think 2 + 3 = 3 + 2 = 5, 1 + 4 = 4 + 1 = 5

If the students have difficulties in doing adding 4, then just let them skip the questions such as 5 + 4. The goal is to master adding 1, 2, and 3 first.

21	32	43	54
31	42	53	42
41	52	42	54
51	31	63	14
41	62		
61	72		
71	32		
41			
81			

12	23	34	45
14	24	35	41
13	14	35	23
14	25	14	
15	26	32	
16	27		
17			
14			
18			

Adding to 6 凑 6

Add digits of each number then write each sum beside that number.
Circle the following numbers when the sum of their digits is 6.

Think 1 + 5 = 5 + 1 = 6, 2 + 4 = 4 + 2 = 6, and double 3 = 6

If the students have difficulties in doing adding 4, then just let them skip the questions such as 5 + 4. The goal is to master adding 1, 2, and 3 first. Some students may have problems to work backwards, in this case just skip to doubles or adding 10 strategies but ask students to try hard before letting them skip. Work with these students on the problem by problem individually.

21	32	43	54
31	42	53	42
51	52	42	
41	62	63	
51	42		
61	72		
71			
51			
81			

12	23	34	45
13	24	35	42
15	25	35	
14	26		
15	24		
16	27		
17	24		
18			

Student Name _____ Date _____

Adding to 7 凑 7

Add digits of each number then write each sum beside that number.
Circle the following numbers when the sum of their digits is **7**.

Think 1 + 6 = 6 + 1, 2 + 5 = 5 + 2, 3 + 4 = 4 + 3, and 1 + 6 = 6 + 1

21	32	43	54
31	52	53	54
61	42	63	52
41	52	43	
51	62		
61	72		
71			
81			

12	23	34	45
13	24	35	46
16	25	35	43
14	26	34	
15	25		
16	27		
17	25		
16			
18			

Student Name _____ Date _____

Adding to 8 凑 8

Add digits of each number then write each sum beside that number.
Circle the following numbers when the sum of their digits is **8**.

Think 1 + 7 = 7 + 1, 2 + 6 = 6 + 2, 3 + 5 = 5 + 3, double 4 = 8

21	32	43	54
31	42	53	43
41	62	63	54
71	52	43	
51	62		
61	72		
71			
81			

12	23	34	45
13	24	35	34
16	25	35	45
14	26	34	
15	27		
16	25		
17			
18			

Adding to 9 凑 9

Add digits of each number then write each sum beside that number.
Circle the following numbers when the sum of their digits is **9**.

Think 1 + 8 = 8 + 1, 2 + 7 = 7 + 2, 3 + 6 = 6, 4 + 5 = 5 + 4 = double 4 + 1

21	32	43	54
31	42	53	53
81	62	53	54
41	52	63	
51	62	53	
8	72		
61			
71			
81			

12	23	34	63
13	24	35	45
14	26	35	35
17	26		
15	27		
17			
18			

Partitioning numbers

Partition 3 拆 3

I have **three** apples and my mom wants me to put them in 2 baskets. How many ways can I put my apples into the two baskets? Write the number of apples in each basket.

Basket A	Basket B

Partition 4 拆 4

I have **four** apples and my mom wants me to put them in 2 baskets. How many ways can I put my apples into the two baskets? Write the number of apples in each basket.

Basket A	Basket B

Partition 5 拆 5

I have **five** apples and my mom wants me to put them in 2 baskets. How many ways can I put my apples into the two baskets? Write the number of apples in each basket.

Basket A	Basket B

Partition 6 拆 6

I have **six** apples and my mom wants me to put them in 2 baskets. How many ways can I put my apples into the two baskets? Write the number of apples in each basket.

Basket A	Basket B

Partition 7 拆 7

I have **seven** apples and my mom wants me to put them in 2 baskets. How many ways can I put my apples into the two baskets? Write the number of apples in each basket.

Basket A	Basket B	Basket A	Basket B

Student Name _____ Date _____

Partition 8 拆 8

I have **eight** apples and my mom wants me to put them in 2 baskets. How many ways can I put my apples into the two baskets? Write the number of apples in each basket.

Basket A	Basket B	Basket A	Basket B

Partition 9 拆 9

I have **nine** apples and my mom wants me to put them in 2 baskets. How many ways can I put my apples into the two baskets? Write the number of apples in each basket.

Basket A	Basket B	Basket A	Basket B

Student Name _____ Date _____

Partition 10 拆 10

I have **ten** apples and my mom wants me to put them in 2 baskets. How many ways can I put my apples into the two baskets? Write the number of apples in each basket.

Basket A	Basket B	Basket A	Basket B

Student Name ______________________________ Date ______________

Number Puzzles

Replace each ? with a number.

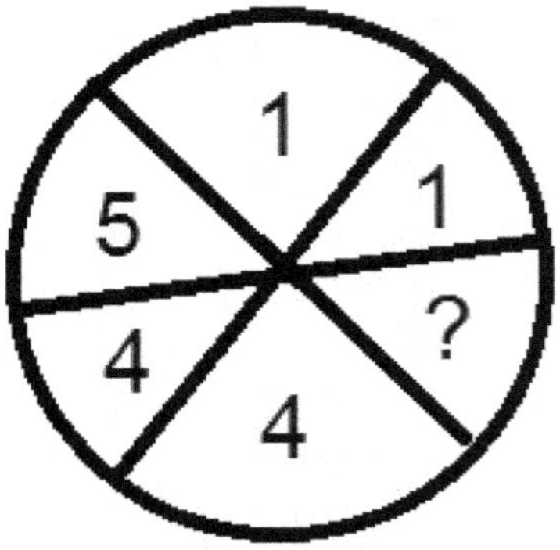

Student Name _____ Date _____

Addition strategy: adding 1, 2, 3 加 1, 2, 3

Adding d + 1 (the larger number in counting)

```
  2 5 3 2 5 4 3 3 4 1 2 3 6 2 7 8 3 4 1 7 3
+ 1 1 1 1 1 1 1 1 1 1 1 1 1 1 1 1 1 1 1 1 1
```

```
  3 6 4 5 7 3 2 1 3 4 5 3 5 1 3 6 3 7 3 8 2
+ 1 1 1 1 1 1 1 1 1 1 1 1 1 1 1 1 1 1 1 1 1
```

```
  5 2 7 1 8 4 3 6 5 4 7 5 2 7 2 4 8 3 6 3 1
+ 1 1 1 1 1 1 1 1 1 1 1 1 1 1 1 1 1 1 1 1 1
```

```
  6 4 5 8 2 7 3 5 7 4 6 3 8 5 3 2 4 8 5 3 7
+ 1 1 1 1 1 1 1 1 1 1 1 1 1 1 1 1 1 1 1 1 1
```

```
  5 7 3 7 4 8 5 3 2 5 7 1 4 6 3 8 4 2 4 8 6
+ 1 1 1 1 1 1 1 1 1 1 1 1 1 1 1 1 1 1 1 1 1
```

Adding d + 1 and 1 + d

```
  0 8 0 8 0 1 0 1 0 1 0 1 0 1 0 1 0 1 0 1 0 1
+ 0 1 0 1 0 7 0 5 0 6 0 3 0 4 0 1 0 2 0 6 0 7
```

```
  0 1 0 1 0 1 0 1 0 1 0 1 0 1 0 1 0 1 0 1 0 1
+ 0 3 0 7 0 4 0 6 0 7 0 8 0 5 0 2 0 3 0 7 0 6
```

```
  0 3 0 1 0 4 0 6 0 7 0 8 0 1 0 3 0 4 0 6 0 6
+ 0 1 0 8 0 1 0 1 0 1 0 1 0 1 0 1 0 1 0 1 0 1
```

```
  0 7 0 8 0 1 0 1 0 1 0 1 0 1 0 1 0 1 0 1 0 1
+ 0 1 0 1 0 7 0 5 0 6 0 3 0 4 0 1 0 2 0 6 0 7
```

```
  0 1 0 1 0 1 0 1 0 1 0 1 0 1 0 1 0 1 0 1 0 1
+ 0 3 0 7 0 4 0 6 0 7 0 8 0 5 0 2 0 3 0 7 0 6
```

2. Add d + 2 and 2 + d.

```
  0 2 0 2 0 2 0 2 0 2 0 2 0 2 0 2 0 2 0 2 0 2
+ 0 3 0 5 0 6 0 7 0 1 0 4 0 5 0 3 0 2 0 5 0 7
```

```
  0 2 0 2 0 2 0 2 0 2 0 2 0 2 0 2 0 2 0 2 0 2
+ 0 6 0 4 0 6 0 7 0 6 0 7 0 2 0 3 0 7 0 6 0 3
```

```
  0 3 0 4 0 6 0 7 0 5 0 1 0 3 0 4 0 6 0 6 0 5
+ 0 2 0 2 0 2 0 2 0 2 0 2 0 2 0 2 0 2 0 2 0 2
```

```
  0 2 0 2 0 2 0 2 0 2 0 2 0 2 0 2 0 2 0 2 0 2
+ 0 7 0 5 0 7 0 5 0 4 0 6 0 4 0 3 0 7 0 6 0 5
```

```
  0 2 0 3 0 7 0 6 0 5 0 4 0 6 0 7 0 6 0 3 0 4
+ 0 2 0 2 0 2 0 2 0 2 0 2 0 2 0 2 0 2 0 2 0 2
```

Adding d + 3 and 3 + d.

```
  0 3 0 3 0 3 0 3 0 3 0 3 0 3 0 3 0 3 0 3 0 3
+ 0 3 0 6 0 1 0 4 0 5 0 3 0 2 0 5 0 4 0 6 0 5
```

```
  0 3 0 3 0 3 0 3 0 3 0 3 0 3 0 3 0 3 0 3 0 3
+ 0 6 0 5 0 4 0 6 0 5 0 5 0 6 0 2 0 3 0 5 0 6
```

```
  0 2 0 6 0 6 0 5 0 1 0 3 0 4 0 6 0 6 0 5 0 3
+ 0 3 0 3 0 3 0 3 0 3 0 3 0 3 0 3 0 3 0 3 0 3
```

```
  0 4 0 6 0 3 0 5 0 2 0 3 0 6 0 4 0 6 0 1 0 1
+ 0 3 0 3 0 3 0 3 0 3 0 3 0 3 0 3 0 3 0 3 0 3
```

```
  0 3 0 6 0 2 0 5 0 5 0 4 0 2 0 5 0 6 0 3 0 5
+ 0 3 0 3 0 3 0 3 0 3 0 3 0 3 0 3 0 3 0 3 0 3
```

```
  0 6 0 3 0 5 0 4 0 6 0 3 0 5 0 3 0 3 0 1 0 3
+ 0 3 0 6 0 3 0 3 0 3 0 4 0 3 0 6 0 5 0 3 0 5
```

d + d with missing operator or number
Fill in each __ with a number or + operator.

6 + 3 = ___	3 + 6 = __
5 + 3= ___	3 + 5 = ___
___ + 3 = 7	__ + 4 = 7
__ + 3 = 6	3 + __ = 6
2 + __ = 5	3 + __ = 5
1 + __ = 4	3 + __ = 4
__ + 3 = 3 + 2 = ___	3 __ 2 = 2 + 3 = ___
3 __ 3 = 2 + 4 = ___	3 __ 3 = 1 + 5 = ___
__ + 3 = 3 + 4 = ___	3 __ 4 = 2 + 5 = ___
5 __ 3 = 2 + 6 = ___	__ + 5 = 1 + 7 = ___
6 + __ = 1 + 8 = ___	3 + __ = 6 + 3 = ___
5 __ 3 = 1 + 7 = ___	3 __ 5 = 2 + 6 = ___
__ + three = 3 + 4 = ___	____ + four = 5 + 2 = ___
five + ____ = 6 + 3 = ___	three + ____ = 6 + 2 = ___
four __ five = 8 + 1 = ___	five __ four = 9 = 7 + 2 = ___

7 days mastering computation skills

After finishing the previous worksheets, the students reach a critical point. The tutor can start to teach or review the following strategies in about 7 days of homework (finish in one week at home or in class.).

The following 7 days of work is very important because it serves as a benchmark for students to see if they could master all addition strategies in a very short time. The tutor must be careful in monitoring a student's progress if after finishing this 7-day of work but still is not proficient in addition strategies. It means that it may take a bit longer for the student than an average student to master these skills, so the tutor must be more patient.

When working on this 7-day of worksheets, the tutor or parent needs to check the student' work after each problem, not to wait until the entire worksheet is finished . Watch how the student works out the solution step by step to make sure that the student understands the reasoning behind each step.

The 7-day strategy short course is only a guide, so if a student could not finish in 7 days, be prepared to give students more time to do it. It is important to test students after they finish at the end of each day to make sure they understand.

If you do not see the students while they are doing it, then you need to teach each day's work before they work on them and make them do a few questions of each day's work, so they do not get stuck while they try to finish them alone.

Day 1 of doubling 1, 2, or 3

Day 1	$1 + 1 = \square$
	$2 + 1 = \square$ $2 + 2 = \square$ = double 2 = $\square$
	$3 + 1 = \square$ $3 + 2 = 2 + \square = \square$ $3 + 3 = \square$ = double 3 = 3 + 3 = $\square$

Day 2 of 2 adding 1, 2, or 3, adding to 10, doubling 2

2 + 2 = 4 = double 2 = 4

d + d with missing operator or number

Fill in each __ with a number or + operator.

2 + 2 = __	2 + 2 = __
2 + 2 = __	2 + 2 = __
__ + 2 = 4	__ + 2 = 4
__ + 2 = 4	__ + 2 = 4
2 + __ = 4	2 + __ = 4
2 + __ = 4	2 + __ = 4
2__ + 2 = 4	2 __ 2 = 4
2 __ 2 = 4	2 __ + 2 = 4
2 __ + 2 = 4	2 __ 2 = 4
2 __ 2 = 4	__ 2 = 4

2 + __ = 4	2 + __ = 4
2 __ 2 = 4	2 __ 2 = 4
__ + two = 4	__ + two = 4
__ + two = 4	__ + two = 4
two __ two = 4	two __ two = 4
two + __ = 4	two __ two= 4
two __ __ = 4	Two __ two = 4
two __ __ =4	two + __ =4

Day 3 of 3 adding 1, 2, or 3, adding to 10, doubling 3

$3 + 3 = 6 = $ double $3 = 6$

d + d with missing operator or number
Fill in each __ with a number or + operator.

3 + 3 = __	3 + 3 = __
3 + 3 = __	3 + 3 = __
__ + 3 = 6	__ + 3 = 6
__ + 3 = 6	__ + 3 = 6
3 + __ = 6	3 + __ = 6
3 + __ = 6	3 + __ = 6
__ + 3 = 6	3 __ 3 = 6
3 __ +3 = 6	__ +3 = 6

Contents include both traditional and Math-Chess-Puzzles combined methods. Extra strength

Student Name _____ Date _____

__　3 = 6	3 __　3 = 6
3 __　3 = 6	__ + 3 = 6
3 + __ = 6	3 + __ = 6
3 __　3 = 6	3 __　3 = 6

__ + three = 6	__ + three = 6
+ three = 6	__ + three = 6
three __　three = 6	__ + three= 6
three + __ = 6	three __　three = 6
three __ __ = 6	three + __ = 6
three __ __ =6	three + __ =6

Day 4 of 4 adding 1, 2, or 3, adding to 10, doubling 4

$4 + 4 = 8 = \text{double } 4 = 8$

d + d with missing operator or number
Fill in each __ with a number or + operator.

4 + 4 =	4 + 4 = __
4 + 4= ___	4 + 4 = ___
___ + 4 = 8	__ + 4 = 8
__ + 4 = 8	__ + 4 = 6
4 + __ = 8	4 + __ = 8
4 + __ = 8	4 + __ = 8
__ + 4 = 8	4 __ 4 = 8
4 __ 4 = 8	__ +4 = 8
__ + 4 = 8	4 __ 4 = 8
4 __ 4 = 8	__ 4 = 8
4 + __ = 8	4 + __ = 8
4 __ 4 = 8	4 __ 4 = 8
__ + four = 8	__ + four = 8
__ + four = 8	four + __ = 8
four __ four = 8	__ four = 8

Student Name _____ Date _____

Day 5 of 5 adding 1, 2, or 3, adding to 10, doubling 5

Day 5			

5 + 1 = 1 + 5 = ☐
5 + 2 = 2 + 5 = ☐
5 + 3 = 3 + 5 = ☐ = double ☐ + 2 = ☐
5 + 4 = 1 + ☐ + 4 = double 4 + ☐ = ☐
5 + 5 = double 5 = ☐

1 + 9 = ☐
2 + 8 = ☐
3 + 7 = ☐
4 + 6 = ☐
5 + 5 = ☐

9 + 1 = ☐
8 + 2 = ☐
7 + 3 = ☐
6 + 4 = ☐
5 + 5 = ☐

Complete the following pattern. For example,

19 = 91

19= ☐
28= ☐
__= ☐
__= ☐
__= ☐

Complete the following pattern. For example,

19 = 91

__= ☐
82= ☐
73= ☐
__= ☐
55= ☐

5 + 5 = 10 = double 5 = 10

d + d with missing operator or number
Fill in each __ with a number or + operator.

5 + 5 =	4 + 4 = __
5 + 5= ___	5 + 5 = ___
___ + 5 = 10	__ + 5 = 10
__ + 5 = 10	__ + 5 = 6

Student Name _____ Date _____

5 + __ = 10	5 + __ = 10
5 + __ = 10	5 + __ = 10
__ + 5 = 10	5 __ 5 = 10
5 __ 5 = 10	__ +5 = 10
__ + 5 = 10	5 __ 5 = 10
5 __ 5 = 10	__ 5 = 10
5 + __ = 10	5 + __ = 10
5 __ 5 = 10	5 __ 5 = 10
__ five = 10	five __ + five = 10
__ + five = 10	five __ + five = 10
five __ five = 10	five __ five = 10

Student Name _____　Date _____

Day 6 of 6 adding 1, 2, or 3, adding to 10, doubling 6

| Day 6 | 6 + 1 = 1 + 6 = ☐
 6 + 2 = 2 + 6 = ☐
 6 + 3 = 3 + 6 = ☐
 6 + 4 = 4 + 6 = ☐
 6 + 5 = 5 + 6 = double 5 + ☐ = ☐
 6 + 6 = double 6 = ☐ | Complete the following pattern.

 19, 28, 37, ___, ____

 ___, 28, 37, ___, ___

 ___, ___, 37, 46, ___

 ___, ___, ____, 46, 55

 _____,28, ___, 46, ____

 ____, ____, 37, ____, 55

 ____, 28, _____, 46, ____

 19, _____, _____, ___, 55 |

6 + 6 = 12 = double 6 = 12

d + d with missing operator or number
Fill in each __ with a number or + operator.

6 + 6 = __	6 + 6 = __
6 + 6 = __	6 + 6 = __
__ + 6 = 12	6 __ + 6 = 12
__ + 6 = 12	__ + 6 = 12
6 + __ = 12	6 + __ = 12
6 + __ = 12	6 __ 6 = 12
__ + 6 = 12	6 __ __ = 12
__ __ 6 = 12	6 __ __ = 12
__ + 6 = 12	__ + 6 = 12
6 __ 6 = 12	__ + 6 = 12
6 __ __ = 12	__ __ 6 = 12
__ __ 6 = 12	__ __ 6 = 12
__ + six = 12	six __ six = 12
six __ six = 12	__ + six = 12
six + __ = 12	six + __ = 12
six __ __ = 12	__ + six = 12
__ + six = 12	six __ six = 12
__ __ six = 12	six __ six = 12

d + d with missing operator or number

Fill in each __ with a number or + operator.

7 + 6 = __	7 + 6 = __
6 + 7 = __	6 + 7 = __
__ + 6 = 13	7 __ + 6 = 13
__ + 7 = 13	__ + 7 = 13
6 + __ = 13	6 + __ = 13
7 + __ = 13	7 __ 6 = 13
__ __ 7 = 13	6 __ __ = 13
__ __ 6 = 13	7 __ __ = 13
__ + 6 = 13	__ + 6 = 13
6 __ 7 = 13	__ + 7 = 13
7 __ __ = 13	__ __ 6 = 13
__ __ 7 = 13	__ __ 7 = 13
__ + seven = 13	six __ seven = 13
seven __ six = 13	__ + six = 13
seven __ __ = 13	seven __ __ = 13
six __ __ = 13	six __ __ = 13
__ __ seven = 13	__ __ seven = 13
__ __ six = 13	__ __ six = 13

Student Name _____ Date _____

d + d with missing operator or number

Fill in each __ with a number or a + operator.

6 + 8 = __	6 + 8 = __
8 + 6 = __	8 + 6 = __
__ 8 = 14	6 __ 8 = 14
__ 6 = 14	__ = 6 = 14
8 + __ = 14	8 + __ = 14
6 + __ = 14	6 + __ = 14
8 __ __ = 14	8 __ __ = 14
6 __ __ = 14	6 __ __ = 14
__ +8 = 14	6 __ 8 = 14
8 __ 6 = 14	__ + 6 = 14
__ __ 6 = 14	__ __ 6 = 14
__ __ 8 = 14	__ __ 8 = 14
__ + eight = 14	six __ eight = 14
__ +six= 14	Eight __ six = 14
__ +8 = 14	six __ __ = 14
__ +6= 14	eight __ __ = 14
__ __ eight = 14	__ __ eight = 14
__ __ six =14	__ __ six =14

Student Name _____ Date _____

Day 6 of 7 adding 1, 2, or 3, adding to 10, doubling 7

Day 6	
7 + 1 = 1 + 7 = ☐ 7 + 2 = 2 + 7 = ☐ 7 + 3 = 3 + 7 = ☐ 7 + 4 = 4 + 7 = ☐ + 1 = ☐ 7 + 5 = 5 + 7 = double 5 + ☐ = ☐ 7 + 6 = double 6 + ☐ = ☐	Complete the following pattern. 91, 82, 73, 64, ___ ___, ___, ___, 64, 55 91, ___, 73, ___, 55 ___, 82, ____, 64, ___ ___, ___, 73, 64, 55 ___, ___, ___, 64, 55 91, ___, ___, ___, 55

7 + 7 = 14 = double 7 = 14

d + d with missing operator or number

Fill in each __ with a number or + operator.

7 + 7 = __	7 + 7 = __
7 + 7 = __	7 + 7 = __
__ + 7 = 14	7 __ + 7 = 14
__ + 7 = 14	__ + 7 = 14
7 + __ = 14	7 + __ = 14
7 + __ = 14	7 __ 7 = 14
__ __ = 14	7 __ __ = 14
__ __ 7 = 14	7 __ __ = 14
__+ 7 = 14	__+ 7 = 14
7 __ 7 = 14	__+ 7 = 14
7 __ __ = 14	__ __ 7 = 14
__ __ 7 = 14	__ __ 7 = 14
seven __ + seven = 14	seven __seven = 14
seven __ seven = 14	__+ seven = 14
seven __ __ seven = 14	seven __ __ seven = 14
seven __ __ = 14	seven __ __ = 14
seven __ __ seven = 14	seven __ __ seven = 14
__ __ seven = 14	__ __ seven = 14

84

Student Name _____ Date _____

Day 7 of 8 adding 1, 2, or 3, adding to 10, doubling 8

Day 7		
	8 + 1 = 1 + 8 = ☐	Complete the following pattern.
	Making 10	91, 82, 73, 64, ___
	8 + 2 = 2 + 8 = ☐	___, ___, ___, 64, 55
	8 + 3 = 3 + 8 = ☐ + 2 + 1 = ☐ + 1 = ☐	91, ___, 73, ___, 55
	8 + 4 = 4 + 8 = ☐ + 2 + 2 = ☐ + 2 = ☐	___, 82, ___, 64, ___
	Doubling	___, ___, 73, 64, 55
	8 + 5 = 5 + 8 = double 5 + ☐ = ☐	___, ___, ___, 64, 55
	8 + 6 = double 6 + ☐ = ☐	91, ___, ___, ___, 55
	8 + 7 = double 7 + ☐ = ☐	
	8 + 8 = double 8 = ☐	

Student Name _____ Date _____

d + d with missing operator or number

Fill in each __ with a number or a + operator.

5 + 8 = __	♖ + 8 = __
8 + ♖ = __	8 + 5= __
__+8 = 13	__ + 8 = 13
__ + 5 = 13	__ + ♖ = 13
8 + __ = 13	8 + __ = 13
♖ + __ = 13	5 + __ = 13
8 __ __ = 13	8 __ __ = 13
5 __ __ = 13	♖ __ __ = 13
♖ __ __ = 13	5 __ __ = 13
__ + 5 = 13	8 __ 5 = 13
__ __ 5= 13	__ __ ♖ = 13
__ __ 8 = 13	__ __ 8 = 13
__ + eight = 13	__ + eight = 13
__ + five = 13	__ + five = 13
five __ __ = 13	five __ __ = 13
eight __ __ = 13	eight __ __ = 13
__ __ eight = 13	__ __ eight = 13
__ __ five =13	__ __ five = 13

d + d with missing operator or number

Fill in each __ with a number or + operator.

7 + 8 = __	7 + 8 = __
8 + 7 = __	8 + 7 = __
__ 8 = 15	__ 8 = 15
__ 7 = 15	__ 7 = 15
8 + __ = 15	8 + __ = 15
7 + __ = 15	7 + __ = 15
__+7 = 15	__+7 = 15
__ +8= 15	__+8 = 15
__+ 8 = 15	__+ 8 = 15
__+ 7 = 15	__+ 7 = 15
8 + __ = 15	8 + __ = 15
7 + __ = 15	7 + __ = 15
__+ eight = 15	__+ eight = 15
__+ seven = 15	__+ seven = 15
__+8 = 15	__ + 8 = 15
__ +7 = 15	__ + 7 = 15
7 + __ = 15	7 + __ = 15
8 + __ s =15	8 + __ s = 15

8 + 8 = 16 = double 8 = 16

d + d with missing operator or number

Fill in each __ with a number or + operator.

8 + 8 = __	7 + 8 = __
8 + 8 = __	8 + 8 = __
__ + 8 = 16	__ + 8 = 16
__ + 8 = 16	__ + 8 = 16
8 + __ = 16	8 + __ = 16
8 + __ = 16	8 + __ = 16
__+8 = 16	__ + 8 = 16
__ +8 = 16	__ + 8 = 16
__+ 8 = 16	__ + 8 = 16
__+ 8 = 16	__+ 8 = 16
8 + __ = 16	8+ __ = 16
8 + __ = 16	8 + __ = 16
__+ eight = 16	__+ eight = 16
__+ eight = 16	__ eight = 16
eight __eight = 16	__ +8= 16
__ + eight = 16	__ + eight = 16
eight + __ = 16	eight __ eight = 16
eight + __ = 16	eight + __ = 16

9 plus strategy

Day 7	
$9 + 1 = 1 + 9 = \square$ $\mathbf{9 + d = 9 + 1 + (d - 1)}$ $9 + 2 = 2 + 9 = 9 + 1 + \square = \square$ $9 + 3 = 3 + 9 = 9 + 1 + \square = \square$ $9 + 4 = 4 + 9 = 9 + 1 + \square = \square$ $9 + 5 = 5 + 9 = 9 + 1 + \square = \square$ $9 + 6 = 6 + 9 = 9 + 1 + \square = \square$ $9 + 7 = 7 + 9 = 9 + 1 + \square = \square$ $9 + 8 = 8 + 9 = 9 + 1 + \square = \square$ $9 + 9 = $ double $9 = \square$	Complete the following pattern. 91, 82, 73, 64, ___ ___, ___, ___, 64, 55 91, ___, 73, ___, 55 ___, 82, ____, 64, ___ ___, ___, 73, 64, 55 ___, ___, ___, 64, 55 91, ___, ___, ___, 55

Student Name _____ Date _____

| Day 7 | $9 + 1 = 1 + 9 = \square$

 $\mathbf{9 + d = 9 + 1 + (d - 1)}$
 $9 + 2 = 2 + 9 = 9 + 1 + \square = \square$
 $9 + 3 = 3 + 9 = 9 + 1 + \square = \square$
 $9 + 4 = 4 + 9 = 9 + 1 + \square = \square$
 $9 + 5 = 5 + 9 = 9 + 1 + \square = \square$
 $9 + 6 = 6 + 9 = 9 + 1 + \square = \square$
 $9 + 7 = 7 + 9 = 9 + 1 + \square = \square$
 $9 + 8 = 8 + 9 = 9 + 1 + \square = \square$

 $9 + 9 = \text{double } 9 = \square$ | Complete the following pattern.

 91, 82, 73, 64, ___

 ___, ___, ___, 64, 55

 91, ___, 73, ___, 55

 ___, 82, ____, 64, ___

 ___, ___, 73, 64, 55

 ___, ___, ___, 64, 55

 91, ___, ___, ___, 55 |

d + d with missing operator or number

Fill in each __ with a number or + operator.

9 + 9 = __	9 + 9 = __
9 + 9 = __	9 + 9 = __
__ + 9 = 18	__ + 9 = 18
__ + 9 = 18	__ + 9 = 18
9 + __ = 18	9 + __ = 18
9 + __ = 18	9 + __ = 18
__ + 9 = 18	__ + 9 = 18
__ + 9 = 18	__ + 9 = 18
__ + 9 = 18	__ + 9 = 18
__ + 9 = 18	__ + 9 = 18
9 + __ = 18	9 + __ = 18
9 + __ = 18	9 + __ = 18
nine __ nine = 18	nine __ nine = 18
__ + nine = 18	__ + nine = 18
nine __ nine = 18	__ + nine = 18
__ + nine = 18	nine __ nine = 18
nine + __ = 18	nine __ nine= 18
nine __ nine = 18	nine + __ = 18

Part 2: More practices of addition strategy : adding 1, 2, 3 加 1, 2, 3

If students are not proficient after finishing Part1, then continue to Part 2 and on.

1 + 1 = _____
2 + 1 = _____
3 + 1 = _____
4 + 1 = _____
5 + 1 = _____
6 + 1 = _____
7 + 1 = _____
8 + 1 = _____

1 + 1 = _____
1+2=2 + 1 = _____
1+3=3 + 1 = _____
1+4 = _____
1+5 = _____
1+6 = _____
1+7 = _____
1+8 = _____

1 + 2 = ___
2 + 2 = ___
3 + 2 = ___
4 + 2 = ___
5 + 2 = ___
6 + 2 = ___
7 + 2 = ___
8 + 2 = ___

Student Name _____ Date _____

1 + 2 = 2 + 1 = _____

2 + 2 = 2 × 2 = _____

2 + 3 = 3 + 2 = _____

2 + 4 = _____

2 + 5 = _____

2 + 6 = _____

2 + 7 = _____

2 + 8 = _____

1 + 2 = 2 + 1 = _____

2 + 2 = 2 × 2 = _____

2 + 3 = 3 + 2 = _____

2 + 4 = _____

2 + 5 = _____

2 + 6 = _____

2 + 7 = _____

2 + 8 = _____

1 + 3 = _____

2 + 3 = _____

3 + 3 = 2 × 3 = _____

4 + 3 = _____

5 + 3 = _____

6 + 3 = _____

7 + 3 = _____

Student Name _____ Date _____

$3 + 1 = 2 + 1 =$ ____

$3 + 2 =$ ____

$3 + 3 = 2 \times 3 =$ ____

$3 + 4 =$ ____

$3 + 5 =$ ____

$3 + 6 =$ ____

$3 + 7 =$ ____

$1 + 4 =$ ____

$2 + 4 =$ ____

$3 + 4 =$ ____

$4 + 4 = 2 \times 4 =$ _____

$5 + 4 =$ ____

$6 + 4 =$ ____

$4 + 1 =$ ____

$4 + 2 =$ ____

$4 + 3 =$ ____

$4 + 4 = 2 \times 4 =$ _____

$4 + 5 =$ ____

$4 + 6 =$ ____

$1 + 5 = 5 + 1 =$ ____

$2 + 5 =$ ____

$3 + 5 =$ ____

$4 + 5 =$ ____

$5 + 5 = 2 \times 5 =$ _____

$5 + 1 =$ ____

$5 + 2 =$ ____

$5 + 3 =$ ____

$5 + 4 =$ ____

$5 + 5 = 2 \times 5 =$ _____

$1 + 5 = 5 + 1 =$ ____

$2 + 5 =$ ____

$3 + 5 =$ ____

$4 + 5 =$ ____

$5 + 5 = 2 \times 5 =$ _____

$5 + 1 =$ ____

$5 + 2 =$ ____

$5 + 3 =$ ____

$5 + 4 =$ ____

$5 + 5 = 2 \times 5 =$ _____

Student Name ______________________________ Date ____________________

1 + 7 = 7 + 1 = ____
2 + 7 = ____
3 + 7 =____

7 + 1 = ____
7 + 2=____
7 + 3=____

1 + 8 = 8 + 1 = ____
2 + 8 = ____

8 + 1 = ____
8 + 2=____

1 + 9 = 9 + 1 = ____

9 + 1 = ____

Student Name _____　Date _____

Adding 1, 2, or 3 by counting

Counting 0, 1 forward d + d

(♔ = 0, ♙ = 1, ♗ = 3, ♘ = 3, ♖ = 5, ♕ = 9)

Do not use fingers to count but recite numbers in brain.

♖	.1	6	7	♕
+ .0	+ ♖	+ .♔	+ .♙	+ .♙
.♙	7	♗	7	♖
+ 8	+ .♙	+ .♙	+ .♔	+ .1
`34				
♖	.♙	.♙	♗	.♕
+ .♙	+ ♕	+ 7	+ .♙	+ 8
♘	♖	♘	.1	♕
+ .1	+ .1	+ .1	+ ♙	+ .1
♗	♖	♗	2	♗
+ .2	+ .1	+ .1	+ ♖	+ .1

Mom! I Learn Addition Using Math-Chess-Puzzles Connection

Contents include both traditional and Math-Chess-Puzzles combined methods. Extra strength

Student Name _____ Date _____

Counting 0, 1 forward d + d (Start with bigger number and count on dots. zero property)
(♔ = 0, ♙ = 1, ♗ = 3, ♘ = 3, ♖ = 5, ♕ = 9)

5	.1	6	7	8
+ ♕	+ 5	+ 0	+ .1	+ .♙
.1	7	6	7	♖
+ 8	+ .1	+ .♙	+ ♕	+ .1
6	♙	6	7	♔
+ .1	+ 9	+ .1	+ .1	+ 8
8	♖	6	7	♕
+ .♙	+ .1	+ .♙	+ .2	+ .♙
♙	7	6	.♙	8
+ 8	+ .1	+ .♙	+ 6	+ .1

Counting forward 2 for d + d with no carrying

(Start with bigger number and count on dots.) (♚ = 0, ♟ = 1, ♝ = 3, ♞ = 3, ♜ = 5, ♛ = 9)

♜	..2	6	..2	7
+ ..2	+ 5	+ ..2	+ 7	+ ..2
..2	7	..2	7	♜
+ 6	+ ..2	+ 4	+ ..2	+ ..2
6	..2	..2	7	..2
+ ..2	+ 7	+ 2	+ ..2	+ 6
3	..2	7	..2	6
+ ..2	+ ♜	+ ..2	+ 7	+ ..2
..2	..2	6	..2	4
+ 5	+ 7	+ ..2	+ 6	+ ..2

Mom! I Learn Addition Using Math-Chess-Puzzles Connection

Contents include both traditional and Math-Chess-Puzzles combined methods. Extra strength

Student Name _____ Date _____

Counting forward 3 for d + d with no carrying
(Start with a bigger number and count on dots.) (♔ = 0, ♙ = 1, ♗ = 3, ♘ = 3, ♖ = 5, ♕ = 9)

♗ + …♗	…♗ + 4	6 +…♗	…♗ + ♖	6 +…♗
…♗ + 6	♖ + …♗	…♗ + 4	4 + …3	♖ + …♗
5 + ..2	…3 + 6	♗ + …3	♖ + …3	…♗ + 6
♗ + …3	…3 + ♖	6 + …3	…♗ + 4	6 + …3
♖ + …♘	4 + …♗	6 + …♘	…♗ + ♖	4 + …♗

Mom! I Learn Addition Using Math-Chess-Puzzles Connection
Contents include both traditional and Math-Chess-Puzzles combined methods. Extra strength
Student Name _____ Date _____

Adding up to 10 by adding 1

5		●		●	
4	●	●	●		●	●
3		●	🐷	●	
2	●	●	●		●	●
1		●		●	
	a	b	c	d	e

You are a chess piece located at c3.
● = 1
Only transfer each number itself to a correct position. Do not the transfer the entire 3 by 3 square. 只放正個別数即可, 不需放正表格.

1	7	4
6	2	8
5	9	3

=__7+1=8

=__1+7=8

7
+
1
——
8

1
+
7
——
8

1
+
8
——
9

=__

=__

=__

=__

Student Name _____ Date _____

Adding up to 10

Only transfer each number itself to a correct position. Do not the transfer the entire 3 by 3 square.

You are a chess piece located at c3.

● = 1

Only transfer each number to a correct position. Do not the transfer the 3 by 3 square.

Adding up to 10

	a	b	c	d	e
5		●		●	
4	●	●	●	●	●
3		●		●	
2	●	●	●	●	●
1		●		●	

Only transfer each number itself to a correct position. Do not the transfer the entire 3 by 3 square.

You are a chess piece located at c3 .
● = 1

Only transfer each number to a correct position. Do not the transfer the 3 by 3 square.

Adding up to 10 by adding 2

You are a chess piece located at c3 [grid].
● = 1

Only transfer each number to a correct position.
Do not the transfer the 3 by 3 square.

Adding up to 10

You are a chess piece located at c3.

● = 1

Only transfer each number to a correct position. Do not the transfer the 3 by 3 square. Each number is transformed into 2 operations.

flip vertically

3	2	4
3	7	8
8	5	1

rotate left 90°

3	2	4
6	7	8
8	5	1

Student Name _____ Date _____

Adding up to 10

You are a chess piece located at c3 .
● = 1

**Only transfer each number to a correct position.
Do not the transfer the 3 by 3 square.**

Student Name _____ Date _____

Adding up to 10 by adding 3

	a	b	c	d	e
5					
4					
3					
2					
1					

You are a chess piece located at c3.
● = 1

Only transfer each number to a correct position.
Do not the transfer the 3 by 3 square.

107

Adding up to 10

You are a chess piece located at c3 ⊞ .

● = 1

Only transfer each number to a correct position.
Do not the transfer the 3 by 3 square.

Adding up to 10

You are a chess piece located at c3.

● = 1

**Only transfer each number to a correct position.
Do not the transfer the 3 by 3 square.**

Student Name _____ Date _____

Addition strategy: Doubling = 2 times (even numbers) 双数

Fill in each ☐ with a number.

Math and Chess sentence	Number sentence.	Comments
♟♟ + 2	2 + 2 = ☐	2 of ♟♟ = 2 × 2 = ☐
♟♟♟ + ♟♟♟	3 + 3 = ☐	2 of ♟♟♟ = 2 × 3 = ☐
♟♟♟♟ + ♟♟♟♟	4 + 4 = ☐	2 of ♟♟♟♟ = 2 × 4 = ☐
♟♟♟♟♟ + ♟♟♟♟♟	5 + 5 = ☐	2 of ♟♟♟♟♟ = 2 × 5 = ☐
6 of ♟ + 6 of ♟	6 + 6 = ☐	2 of 6 ♟ = 2 × 6 = ☐
7 of ♟ + 7 of ♟	7 + 7 = ☐	2 of 7 ♟ = 2 × 7 = ☐
8 of ♟ + 8 of ♟	8 + 8 = ☐	2 of 8 ♟ = 2 × 8 = ☐
9 of ♟s + 9 of ♟	9 + 9 = ☐	2 of 9 ♟ = 2 × 9 = ☐

Getting even numbers by doubling a number

It is an easy task for some students to understand the concepts of adding 1, 2, 3 or adding to ten or doubling but for some students who sit in the same math class, the work to come up with a correct answer becomes extremely difficult or slow. So here it is we have one more review of doubling concept.

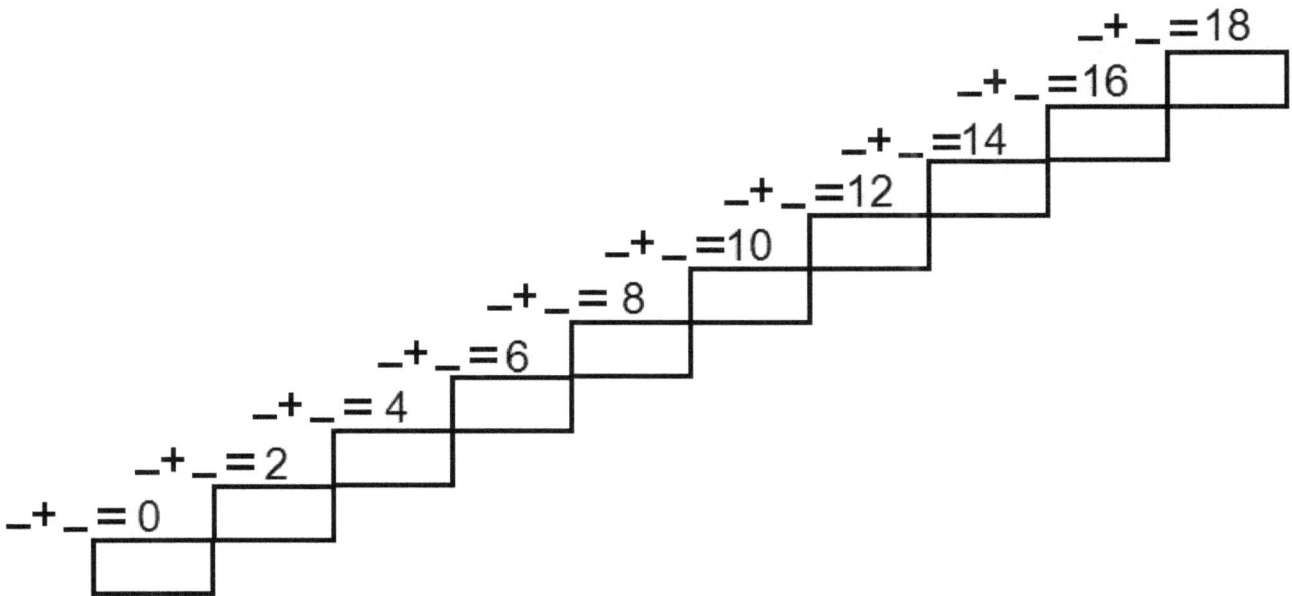

_+_=18
_+_=16
_+_=14
_+_=12
_+_=10
_+_= 8
_+_= 6
_+_= 4
_+_= 2
_+_= 0

Doubling = 2 times (even numbers)

This page is not meant to teach multiplication but shows the relationships of doubling.

2 + 2 = ____	2 + 2 = 2 of 2 = ☐ × 2 =	2 plus ___ equals 2 times 2 = ___
3 + 3 = ____	3 + 3 = 2 of 3 = ☐ × 3 =	3 plus ___ equals 2 times 3 = ___
4 + 4 = ____	4 + 4 = 2 of 4 = ☐ × 4 =	4 plus ___ equals 2 times 4 = ___
5 + 5 = ____	5 + 5 = 2 of 5 = ☐ × 5 =	5 plus 5 equals ___ times 5 = ___
6 + 6 = ____	6 + 6 = 2 of 6 = ☐ × 6 =	6 plus ___ equals 2 times ___ = ___
7 + 7 = ____	7 + 7 = 2 of 7 = ☐ × 7 =	7 plus 7 equals 2 times ___ = ___
8 + 8 = ____	8 + 8 = 2 of 8 = ☐ × 8 =	8 plus 8 equals 2 times ___ = ___
9 + 9 = ____	9 + 9 = 2 of 9 = ☐ × 9 =	9 plus ___ equals 2 times 9 = ___

Doubling

2	7	4	8	6
+ 2	+ 7	+ 4	+ 8	+ 6
___	___	___	___	___
♗	8	♕	4	♖
+ ♗	+ 8	+ ♕	+ 4	+ ♖
___	___	___	___	___
7	8	9	5	3
+7	+ 8	+ 9	+ 5	+3
___	___	___	___	___
2	7	♕	4	6
+ 2	+ 7	+ ♕	+ 4	+ 6
___	___	___	___	___

Doubling

For each question, fill in each ☐ with the same number.

☐2	☐	☐	☐	☐
+☐2	+☐	+☐	+☐	+☐
4	4	4	4	4
☐3	☐	☐	☐	☐
+☐3	+☐	+☐	+☐	+☐
6	6	6	6	6
☐4	☐	☐	☐	☐
+☐4	+☐	+☐	+☐	+☐
8	8	8	8	8
☐5	☐	☐	☐	☐
+☐5	+☐	+☐	+☐	+☐
10	10	10	10	10

Doubling

For each question, fill in each ☐ with the same number.

☐6	☐	☐	☐	☐
+ ☐6	+ ☐	+ ☐	+ ☐	+ ☐
——	——	——	——	——
☐7	☐	☐	☐	☐
+ ☐7	+ ☐	+ ☐	+ ☐	+ ☐
——	——	——	——	——
☐8	☐	☐	☐	☐
+ ☐8	+ ☐	+ ☐	+ ☐	+ ☐
——	——	——	——	——
☐9	☐	☐	☐	☐
+ ☐9	+ ☐	+ ☐	+ ☐	+ ☐
——	——	——	——	——

Doubling

two	two	three	three	four
+ two	+ two	+ three	+ three	+ four
four	five	five	six	six
+ four	+ five	+ five	+ six	+ six
Seven	seven	eight	eight	nine
+ seven	+ seven	+ eight	+ eight	+ nine
nine	seven	five	four	nine
+ nine	+ seven	+ five	+ four	+ nine
seven	eight	six	seven	six
+ seven	+ eight	+ six	+ seven	+ six

Doubling

two plus two is □	2 + 2 = □
three plus three is □	3 + 3 = □
four plus four is □	4 + 4 = □
five plus five is □	5 + 5 = □
six plus six is □	6 + 6 = □
seven plus seven is □	7 + 7 = □
eight plus eight is □	8 + 8 = □
nine plus nine is □	9 + 9 = □
two plus two is □	2 + 2 = □
three plus three is □	3 + 3 = □
four plus four is □	4 + 4 = □
five plus five is □	5 + 5 = □
six plus six is □	6 + 6 = □
seven plus seven is □	7 + 7 = □
eight plus eight is □	8 + 8 = □
nine plus nine is □	9 + 9 = □

d + d with missing operator or number

Fill in each __ with a number or + operator.

9 + 9 = __	9 + 9 = __
9 + 9 = __	9 + 9 = __
__ + 9 = 18	__ + 9 = 18
__ + 9 = 18	__ + 9 = 18
9 + __ = 18	9 + __ = 18
9 + __ = 18	9 + __ = 18
__+ 9 = 18	__+ 9 = 18
__ + 9 = 18	__+ 9 = 18
__+ 9 = 18	__+ 9 = 18
__+ 9 = 18	__+ 9 = 18
9 + __ = 18	9 + __ = 18
9 + __ = 18	9 + __ = 18
nine __ nine = 18	nine __ nine = 18
__+ nine = 18	__ + nine = 18
nine __ nine = 18	__ + nine = 18
__ + nine = 18	nine __ nine = 18
nine + __ = 18	nine __ nine= 18
nine __ nine = 18	nine + __ = 18

Working backwards on doubling

$\bigcirc + 0 = 0$	$4 + \bigcirc = 8$
$4 + \bigcirc = 8$	$\bigcirc + 0 = 0$
$\bigcirc + 2 = 4$	$6 + \bigcirc = 12$
$6 + \bigcirc = 12$	$\bigcirc + 0 = 0$
$\bigcirc + 9 = 18$	$5 + \bigcirc = 10$
$5 + \bigcirc = 10$	$\bigcirc + 7 = 14$
$\bigcirc + 3 = 6$	$8 + \bigcirc = 16$
$2 + \bigcirc = 4$	$\bigcirc + 0 = 0$
$\bigcirc + 1 = 2$	$6 + \bigcirc = 12$
$4 + \bigcirc = 8$	$\bigcirc + 0 = 0$
$\bigcirc + 2 = 4$	$4 + \bigcirc = 8$
$7 + \bigcirc = 14$	$\bigcirc + 0 = 0$
$\bigcirc + 9 = 18$	$5 + \bigcirc = 10$
$5 + \bigcirc = 10$	$\bigcirc + 9 = 18$
$\bigcirc + 3 = 6$	$8 + \bigcirc = 16$
$2 + \bigcirc = 4$	$\bigcirc + 0 = 0$
$\bigcirc + 1 = 2$	$7 + \bigcirc = 14$

Ho Math Chess 何数棋谜　妈! 我会棋谜式加法啦!

Mom! I Learn Addition Using Math-Chess-Puzzles Connection

Contents include both traditional and Math-Chess-Puzzles combined methods. Extra strength

Student Name _____ Date _____

$$\triangle + \triangle - 1 = 7$$

$$\triangle + \triangle + 1 = 9$$

$$\triangle + \triangle - 1 = 13$$

$$\triangle + \triangle + 1 = 15$$

$$\triangle + \triangle - 1 = 11$$

$$\triangle + \triangle + 1 = 13$$

$$\triangle + \triangle - 1 = 17$$

$$\triangle + \triangle + 1 = 19$$

$$\triangle + \triangle - 1 = 3$$

$$\triangle + \triangle + 1 = 7$$

$$\triangle + \triangle - 1 = 19$$

$$\triangle + \triangle + 1 = 23$$

$$\triangle + \triangle - 1 = 13$$

$$\triangle + \triangle + 1 = 21$$

$$\triangle + \triangle - 1 = 1$$

$$\triangle + \triangle + 1 = 3$$

Student Name _____ Date _____

$2 + 4 + \bigcirc = 9$	$\bigcirc + \bigcirc + 5 = 9$
$2 + 2 + \bigcirc = 6$	$\bigcirc - 1 = 8$
$\bigcirc + \bigcirc + \bigcirc + 1 = 10$	$\bigcirc + \bigcirc = 8$
$\bigcirc + \bigcirc + 4 = 8$	$\bigcirc + \bigcirc - 1 = 9$
$\bigcirc + \bigcirc + 1 = 7$	$\bigcirc + \bigcirc - 2 = 8$
$\bigcirc + \bigcirc + 2 = 8$	$\bigcirc + \bigcirc + 2 = 8$
$8 + \bigcirc = 9$	$\bigcirc + \bigcirc - 1 = 7$
$8 + \bigcirc + \bigcirc = 10$	$\bigcirc + \bigcirc + 1 = 7$
$\bigcirc + \bigcirc + \bigcirc = 9$	$\bigcirc + \bigcirc + 3 = 7$
$7 + \bigcirc + 1 = 10$	$\bigcirc + \bigcirc + 5 = 9$
$5 + 2 + \bigcirc = 8$	$\bigcirc + \bigcirc - 1 = 9$
$4 + \bigcirc + \bigcirc = 6$	$\bigcirc + \bigcirc + 2 = 10$
$3 + \bigcirc + \bigcirc = 7$	$\bigcirc + \bigcirc - 2 = 10$
$\bigcirc + \bigcirc + 1 = 5$	$\bigcirc + \bigcirc + 3 = 11$
$\bigcirc + \bigcirc - 1 = 5$	$\bigcirc + \bigcirc - 1 = 11$
$\bigcirc + 3 = 6$	$8 + \bigcirc = 16$
$2 + \bigcirc = 4$	$\bigcirc + 0 = 0$
$\bigcirc + 1 = 2$	$7 + \bigcirc = 14$

Student Name _____　Date _____

$\bigcirc + \bigcirc + 1 = 5$

$\bigcirc + \bigcirc + 2 = 10$

$\bigcirc + \bigcirc + 1 = 5$

$\bigcirc + \bigcirc + 9+1 = 20$

$4 + \bigcirc + \bigcirc + 6 = 14$

$7 + 3 + \bigcirc + \bigcirc + 1 = 23$

$9 + \bigcirc + \bigcirc + 1 = 18$

$\triangle + \triangle = 8,\ \triangle + \bigcirc + \bigcirc + 2 = 16$

$\bigcirc + \bigcirc - 1 = 13$

$\bigcirc + \bigcirc + 11 - 1 = 28$

$4 + \bigcirc + \bigcirc - 6 = 14$

$7 - 3 + \bigcirc + \bigcirc - 1 = 17$

$\bigcirc + 2 + \triangle = \triangle + 3 + 1$

$11 - 5 = \triangle$

$\triangle + \triangle = \bigcirc + \bigcirc + 4$

Student Name _____ Date _____

$1 + 1 + \bigcirc = 4$

$3 + 1 + \bigcirc = 8$

$1 + 5 + \bigcirc = 12$

$7 + \bigcirc - 1 = 14$

$9 + \bigcirc - 1 = 18$

$9 - \bigcirc + 8 = 16$

$7 + \bigcirc + 8 = 16$

$9 + 10 - \bigcirc = 18$

$4 - \bigcirc + 3 = 6$

$\bigcirc - 1 + 5 = 10$

$9 + 1 + 6 + \bigcirc = 20$

$8 + 2 + 7 + \bigcirc = 20$

$5 + 5 + 5 + \bigcirc = 20$

$4 + 6 + 3 + \bigcirc = 20$

$2 + 1 + 7 + 2 + 2 + \bigcirc = 20$

$10 + \bigcirc + 9 = 20$

$3 + 7 + \bigcirc + 6 = 20$

$5 + \bigcirc + 3 + 7 = 20$

If $\triangle + 3 = 9$, then $\triangle + \triangle + \bigcirc = 16$. What is the value of $\bigcirc$?

$\bigcirc + 3 = \bigcirc + 2 + ?$

$\bigcirc + \bigcirc + 2 = 8$, What is the value of $\bigcirc$?

Doubling by relating numbers

8 + 9 uses the double of the smaller number 8. 6 + 7 uses the double of the smaller number 6.

8 + 8 = _____ 8 + 9 = ☐ +1 + 9 = ___ 9 − 2 + 8 = __ ____	5 + 5 = _____ 5 + 6 = _____ = 5 + ☐ +1= 6 − 2 + 5 = _____
6+6 = _____ 6 + 7 = _____ = 6 + ☐ + 1 = _ 8−2+7= _____	9 + 9 = _____ 9 + 8 = 9 + 1 + ☐ = _____ 10 − 2 + 9 = _____
7 + 7 = _____ 7 + 8 = 7 + ☐ +1 = _____ 9 − 2 + 8 = _____	3 + 3 = _____ 3 + 4 = _____ 5 − 2 + 4 = _____
4 + 4= _____ 4 + 5 = _____ 6 − 2 + 5 = _____	6 + 6 = _____ 7 + 6 =1+ ☐ + 6 = _____ 8 + 7 = _____ 9 + 7 = _____

Student Name _____ Date _____

Using doubles to add

(♔ = 0, ♙ = 1, ♗ = 3, ♘ = 3, ♖ = 5, ♕ = 9)

♙	10	2	20	3
+ ♙	+ 10	+ 2	+ 20	+ 3
♞	30	4	40	♖
+ ♞	+ 30	+ 4	+ 40	+ ♖
50	6	60	7	70
+ 50	+ 6	+ 60	+ 7	+ 70
8	80	9	90	10
+ 8	+ 80	+ 9	+ 90	+ 10
20	30	40	50	60
+ 20	+ 30	+ 40	+ 50	+ 60

Ho **Math Chess** 何数棋谜　妈！我会棋谜式加法啦！
Mom! I Learn Addition Using Math-Chess-Puzzles Connection
Contents include both traditional and Math-Chess-Puzzles combined methods. Extra strength
©2008 – 2018 Frank Ho, Amanda Ho　　All rights reserved. www.homathchess.com

Student Name _____ Date _____

Doubling

3	4	6	8
2	10	✳	12
1	14	16	18
	a	b	c

⊢→⊢→⊢ No
⊢→⊢ Yes

Student Name _____ Date _____

($\text{♔} = 0$, $\text{♙} = 1$, $\text{♗} = 3$, $\text{♘} = 3$, $\text{♖} = 5$, $\text{♕} = 9$)

$4 + \square = 8$	$\square + 4 = 8$	$4 + \square = 8$	$\square + 4 = 8$
$\square + 4 = 8$	$4 + \square = 8$	$\square + 4 = 8$	$4 \,\square\, 4 = 8$

$4 + 4 = 8$ four + four = $\square$

$$\begin{array}{r} 4 \\ + 4 \\ \hline \square \end{array} \qquad \begin{array}{r} 4 \\ + 4 \\ \hline \square \end{array}$$

$4 + \text{♖} = 4 + 4 + 1$
$\quad = \square + 1$
$\quad = \square$

four + five = $\square$

$$\begin{array}{r} 4 \\ + \text{♖} \\ \hline \square \end{array} \qquad \begin{array}{r} 4 \\ 4 \\ + 1 \\ \hline \square \end{array}$$

$4 + 7 = 1 + 3 + 7$
$\quad = 1 + \square$
$\quad = \square$

Four + seven = $\square$

$$\begin{array}{r} 4 \\ + 7 \\ \hline \end{array} \qquad \begin{array}{r} 1 \\ \text{♗} \\ + 7 \\ \hline \end{array}$$

$4 + \square = 12$	$\square + 8 = 12$	$8 + \square = 12$	$\square + 8 = 12$
$\square + 8 = 12$	$8 + \square = 12$	$\square + 8 = 12$	$8 + \square = 12$

d + d
(♔ = 0, ♙ = 1, ♗ = 3, ♘ = 3, ♖ = 5, ♕ = 9)

5 + □ = 10	□ + 5 = 10	5 + □ = 10	□ + 5 = 10
□ + 6 = 11	5 + □ = 11	□ + 6 = 11	5 + □ = 11

5 + 7 =   five + seven = □

$$\begin{array}{r} ♖ \\ + \ 7 \\ \hline \end{array} \qquad \begin{array}{r} 7 \\ + \ ♖ \\ \hline \end{array}$$

5 + 8 = 5 + 5 + 3   five + eight = □

   = □ + 3

   = □

$$\begin{array}{r} ♖ \\ + \ 8 \\ \hline \end{array} \qquad \begin{array}{r} ♖ \\ ♖ \\ + \ ♗ \\ \hline \end{array}$$

5 + 9 = 4 + ___ + 9   five + nine = □

   = □ 9 + d = 9 +

1+ (d-1)

$$\begin{array}{r} ♖ \\ + \ 9 \\ \hline \end{array} \qquad \begin{array}{r} 4 \\ 1 \\ + \ 9 \\ \hline \end{array}$$

9 + □ = 14	□ + 5 = 14	9 + □ = 14	□ + 5 = 14
□ + 9 = 15	5 + □ = 14	□ + 9 = 14	5 + □ = 14

d + d

6 + □ = 12	□ + 6 = 12	6 + □= 12	□ + 6 = 12
□ + 7 = 13	7 + □ = 13	□ + 7 = 13	7 + □ = 13

6 + 8 =   six + eight = □

$$\begin{array}{r} 6 \\ + \ 8 \\ \hline \end{array}$$ $$\begin{array}{r} 8 \\ + \ 6 \\ \hline \end{array}$$

8 + 6 = 6 + 6 + 2   eight + six = □
    =□ + 2
    = □

$$\begin{array}{r} 8 \\ + \ 6 \\ \hline \end{array}$$ $$\begin{array}{r} 2 \\ 6 \\ + \ 6 \\ \hline \end{array}$$

6 + 9 = 1 + 5 + 9   six + nine = □
    = 10 + □
    = □ Split the
small number to be 1
and a smaller
number.

$$\begin{array}{r} 6 \\ + \ 9 \\ \hline \end{array}$$ $$\begin{array}{r} 1 \\ 5 \\ + \ 9 \\ \hline \end{array}$$

9 + □ = 15	□ + 6 = 15	9 + □ = 15	□ + 6 = 15
□ + 8 = 14	8 + □ = 14	□ + 8 = 14	8 + □ = 14

d + d

7 + □ = 14	□ + 7 = 14	7 □7 = 14	□ + 7 = 14
□ + 7 = 14	7 + □ = 14	□+ 7 = 14	7 □7 = 14

7 + 7 =   seven + seven = □

$$7 \atop +\,7$$ $$7 \atop +\,7$$

7 + 8 = 7 + 7 + 1   seven + eight = □15
   = □ 14+ 1
   = □ 15

$$7 \atop +\,8$$ $$7 \atop 7 \atop +\,♙$$

8 + 7 = 1 + 7 + 7   eight + seven = □15
   = 1 + □14
   = □ 15

$$8 \atop +\,7$$ $$♙ \atop 7 \atop +\,7$$

7 +□ = 15	□ + 7 = 15	7 + □= 14	□ + 7 = 15
□ + 8 = 15	8 + □ = 14	□ + 8 = 14	8 + □ = 14

Student Name _____ Date _____

d + d

8 + □ = 16	8□ 8 = 16	8 + □ = 16	□ + 8 = 16
□ + 9 = 17	8 + □ = 17	□ + 9 = 17	8 + □= 17

8 + 8 =   eight + eight = □

$$\begin{array}{r} 8 \\ + 8 \\ \hline \end{array} \qquad \begin{array}{r} 8 \\ + 8 \\ \hline \end{array}$$

8 + 9 = 7 + 1 + 9   seven + nine = □16
    = □ + 10
    = □

$$\begin{array}{r} 8 \\ + ♛ \\ \hline \end{array} \qquad \begin{array}{r} 1 \\ 7 \\ + ♛ \\ \hline \end{array}$$

9 + 8 = 9 + 1 + 7   nine + eight = □
    = 10 + □
    = □

$$\begin{array}{r} ♛ \\ + 8 \\ \hline \end{array} \qquad \begin{array}{r} ♛ \\ 1 \\ + 7 \\ \hline \end{array}$$

8 + □ = 17	□ + 9 = 17	8 + □ = 17	□ + 9 = 17
□ + 9 = 17	8 + □ = 17	□ + 9 = 17	8 + □ = 17

Ho Math Chess 何数棋谜　妈！我会棋谜式加法啦！

Mom! I Learn Addition Using Math-Chess-Puzzles Connection

Contents include both traditional and Math-Chess-Puzzles combined methods. Extra strength

©2008 – 2018 Frank Ho, Amanda Ho　　All rights reserved. www.homathchess.com

Student Name _____ Date _____

d + d

7 + 7	7 + 7	7 + 7	7 + 8	8 + 7
7 + 8	8 + 7	7 + 8	8 + 7	8 + 7
8 + 7	7 + 8	8 + 7	7 + 8	8 + 7
7 + 8	8 + 7	7 + 8	8 + 7	8 + 7

Student Name _____ Date _____

d + d

6 + 6	7 + 6	6 + 7	7 + 6	6 + 7
7 + 6	6 + 7	7 + 6	6 + 7	6 + 7
6 + ♛	6 + 9	9 + 6	♛ + 6	6 + ♛
8 + 6	6 + 8	6 + 8	8 + 6	6 + 8

Student Name ______________________________ Date ____________________

d + d

8	8	8	8	8
+ 8	+ 8	+ 8	+ 8	+ 8
8	9	♕	8	8
+ 9	+ 8	+ 8	+ ♕	+ ♕
8	♕	8	9	8
+ ♕	+ 8	+ ♕	+ 8	+ 9
♕	8	♕	8	9
+ 8	+ 9	+ 8	+ ♕	+ 8

Student Name ______________________________ Date ____________________

d + d

6	6	6	7	6
+ 6	+ 6	+ 6	+ 6	+ 7
7	6	7	6	6
+ 6	+ 7	+ 6	+ 7	+ 7
8	6	6	6	8
+ 6	+ 8	+ 8	+ 8	+ 6
♕	6	♕	6	6
+ 6	+ 9	+ 6	+ ♕	+ 9

Student Name ______________________________ Date ____________________

d + d

♖	5	5	6	♖
+ ♖	+ 5	+ 6	+ ♖	+ 6
8	5	♖	8	5
+ 5	+ 8	+ 8	+ 5	+ 8
7	7	5	♖	7
+ 5	+ 5	+ 7	+ 7	+ ♖
♖	9	♖	9	5
+ 9	+ 5	+ 9	+ ♖	+ 9

d + d with missing operator or number

Fill in □ with a number or + operator

7 + 7 = □	7 + 7 = □
7 + 7 = □	7 + 7 = □
□ + 7 = 14	□ + 7 = 14
□ + 7 = 14	□ + 7 = 14
7 + □ = 14	7 + □ = 14
7 + □ = 14	7 + □ = 14
□+ 7 = 14	7 □ 7 = 14
7 □ 7 = 14	□+7 = 14
□+ 7 = 14	7 □ 7 = 14
7 □ 7 = 14	□+ 7 = 14
7+□ = 14	7+□ = 14
7□ 7 = 14	6□ 8 = 14
□+ seven = 14	□+ seven = 14
□+ seven = 14	□+ seven = 14
seven □ 7 = 14	seven □7 = 14
seven +□ = 14	seven +□= 14
7□□ = 14	7+□ = 14
7□□ =14	7+□ =14

d + d with missing operator or number

Fill in □ with a number or + operator

7 + 8 = □	7 + 8 = □
8 + 7 = □	8 + 7 = □
□ + 8 = 15	□ + 8 = 15
□ + 7 = 15	□ + 7 = 15
8 + □ = 15	8 + □ = 15
7 + □ = 15	7 + □ = 15
□+7 = 15	□+7 = 15
□ +8= 15	□+8 = 15
□+ 8 = 15	□+ 8 = 15
□+ 7 = 15	□+ 7 = 15
8+□ = 15	8+□ = 15
7+□ = 15	7+□ = 15
□+ eight = 15	□+ eight = 15
□+ seven = 15	□+ seven = 15
□+8 = 15	□ +8= 15
□ +7 = 15	□ +7= 15
7+□ = 15	7+□ = 15
8+□ =15	8+□ =15

d + d with missing operator or number

Fill in □ with a number or + operator

7 + 6 = □	7 + 6 = □
6 + 7 = □	6 + 7 = □
□ + 6 = 13	7□ + 6 = 13
□ + 7 = 13	□ + 7 = 13
6 + □ = 13	6 + □ = 13
7 + □ 13	7 □ 6 = 13
□□7 = 13	6 □□ = 13
□□ 6= 13	7 □□= 13
□+ 6 = 13	□+ 6 = 13
6 □ 7 = 13	□+ 7 = 13
7□□ = 13	□□ 6 = 13
□□ 7 = 13	□□ 7 = 13
□ + seven = 13	six □seven = 13
seven □ six = 13	□+ six = 13
seven □ □ = 13	seven □ □ = 13
six □ □ = 13	six □ □ = 13
□□ seven = 13	□□ seven = 13
□□ six=13	□□ six=13

d + d with missing operator or number

Fill in □ with a number or a + operator

6 + 8 = □	6 + 8 = □
8 + 6 = □	8 + 6 = □
□ + 8 = 14	6□ 8 = 14
□ + 6 = 14	□ + 6 = 14
8 + □ = 14	8 + □ = 14
6 + □ = 14	6 + □ = 14
8 □□ = 14	8 □ □ = 14
6 □□ = 14	6 □□ = 14
□ +8 = 14	6 □ 8 = 14
8 □ 6 = 14	□ + 6 = 14
□□ 6 = 14	□□ 6 = 14
□□ 8 = 14	□□ 8 = 14
□ + eight = 14	six □ eight = 14
□ +six= 14	Eight □ six = 14
□ +8 = 14	six □ □ = 14
□ +6= 14	eight □ □ = 14
□□ eight = 14	□□ eight = 14
□□ six =14	□□ six =14

d + d with missing operator or number

Fill in □ with a number or a + operator

5 + 8 = □	♖ + 8 = □
8 + ♖ = □	8 + 5 = □
□ + 8 = 13	□ + 8 = 13
□ + 5 = 13	□ + ♖ = 13
8 + □ = 13	8 + □ = 13
♖ + □ = 13	5 + □ = 13
8 □□ = 13	8 □□ = 13
5 □□ = 13	♖ □□ = 13
♖ □□ = 13	5 □□ = 13
□ + 5 = 13	8 □ 5 = 13
□□ 5 = 13	□□ ♖ = 13
□□ 8 = 13	□□ 8 = 13
□ + eight = 13	□ + eight = 13
□ + five = 13	□ + five = 13
five □ □ = 13	five □ □ = 13
eight □ □ = 13	eight □ □ = 13
□□ eight = 13	□□ eight = 13
□□ five = 13	□□ five = 13

Student Name _____ Date _____

Doubling, up, and down

5					
4	8	9	8	7	6
3		5	5	5	
2	3	4	3	2	1
1					
	a	b	c	d	e

You are at c3 = ☐ (fill in a number.).

1. ☐ + = ____

2. ☐ + = ____

3. ☐ + = ____

4. ☐ + = ____

5. ☐ + = ____

6. ☐ + = ____

7. ☐ + = ____

8. ☐ + = ____

9. ☐ + = ____

10. ☐ + = ____

11. ☐ + = ____

12. ☐ + = ____

Student Name _____ Date _____

Doubling, up, and down

5					
4	8	9	8	7	9
3		6	6	6	
2	5	4	3	2	1
1					
	a	b	c	d	e

You are at c3 = ☐ (fill in a number.).

1. ☐ + = ____

2. ☐ + = ____

3. ☐ + = ____

4. ☐ + = ____

5. ☐ + = ____

6. ☐ + = ____

7. ☐ + = ____

8. ☐ + = ____

9. ☐ + = ____

10. ☐ + = ____

11. ☐ + = ____

12. ☐ + = ____

Student Name _____ Date _____

Doubling, up, and down

5					
4	9	8	9	8	9
3		7	7	7	
2	6	5	4	3	2
1					
	a	b	c	d	e

You are at c3 = ☐ (fill in a number.).

1. ☐ + = ____

2. ☐ + = ____

3. ☐ + = ____

4. ☐ + = ____

5. ☐ + = ____

6. ☐ + = ____

7. ☐ + = ____

8. ☐ + = ____

9. ☐ + = ____

10. ☐ + = ____

11. ☐ + = ____

12. ☐ + = ____

Doubling

5	4		7		2
4		4	7	2	
3	9	9	1	8	8
2		3	6	5	
1	3		6		5
	a	b	c	d	e

You are a chess piece located at c3.

⤢ + ⤢ = ____ + ____ = ____

⤢ + ⤢ = ____ + ____ = ____

Ho Math Chess 何数棋谜　妈！我会棋谜式加法啦！

Mom! I Learn Addition Using Math-Chess-Puzzles Connection

Contents include both traditional and Math-Chess-Puzzles combined methods. Extra strength

©2008 – 2018 Frank Ho, Amanda Ho　All rights reserved. www.homathchess.com

Student Name _____ Date _____

Doubling

5	4		7		2
4		4	7	2	
3	9	9	1	8	8
2		3	6	5	
1	3		6		5
	a	b	c	d	e

You are a chess piece located at c3.

⟋⟍ + ⟋⟍ = _____ + _____ = _____

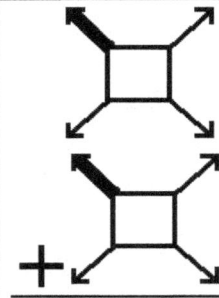

⟋⟍ + ⟋⟍ = _____ + _____ = _____

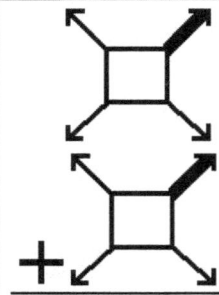

Student Name _____ Date _____

Doubling

5	4		7		2
4		4	7	2	
3	9	9	1	8	8
2		3	6	5	
1	3		6		5
	a	b	c	d	e

You are a chess piece located at c3.

⬍ + ⬍ = ____ + ____ = ____

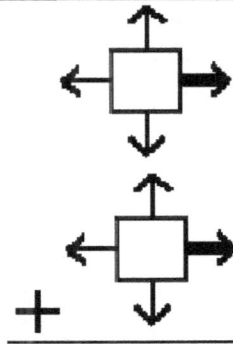

⬍ + ⬍ = ____ + ____ = ____

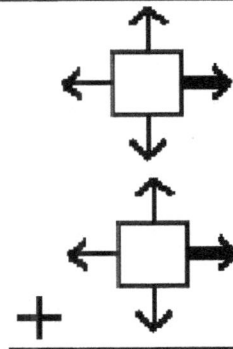

Adding strategy: adding to 10 凑十

Memorizing the following making up to 10. Replace each ? with a digit.

Remember 1 +9 = 2 + 8 = 3 + 7 = 4 + 6 = 5 + 5 = 10 (19, 28, 37, 46, 55)

For each column, the left digit + the right digit = 10	
9 1	**1 9**
8 2	**2 8**
7 3	**3 7**
6 4	**4 6**
5 5	**5 5**

9 1	1 9
8 2	? ?
7 3	3 7
6 4	? ?
5 5	? ?

9 1	? ?
8 2	2 ?
7 3	7 ?
6 4	? ?
5 5	? ?

9 1	? ?
? ?	2 8
7 3	? ?
? ?	4 6
? ?	5 5

? ?	1 9
8 2	? ?
? ?	3 7
6 4	? ?
? ?	5 5

9 1	1 ?
8 2	2 ?
7 3	? 7
6 4	? 6
5 5	? 5

? 1	1 ?
8 ?	? 8
? 3	3 ?
6 ?	? 6
? 5	5 ?

? ?	? 9
8 ?	? ?
? 3	3 ?
? ?	? 6
5 5	? ?

9 ?	? ?
? 2	2 ?
7 ?	? ?
6 4	? 6
? 5	5 ?

Counting a pair of numbers to make 10.

9 8 7 6 5 5 3 2 1 4
1 2 3 4 5 5 6 7 8 9
5 4 6 7 8 3 2 1 9 5
7 4 3 2 1 9 6 5 8 5
8 3 4 6 5 2 1 9 7 5
4 5 7 6 3 1 9 2 8 5
3 6 8 7 2 9 1 5 4
8+1, 4+1, 7+1, 5+1, 6+1,2+1, 1+1, 0+1, 4+1, 3+1
6-1, 9, 8-1, 9-1, 6-1, 7-1, 4-1, 5-1, 3-1, 1
2-1, 7+1, 4-1, 3+1, 6-1, 1+1, 8-1, 5+1, 8+1, 6-1

Student Name _____ Date _____

Circle all tens.

9 + 1=	5 + 5=	9 + 2=
8 + 3=	4 + 6=	2 + 8=
7 + 4=	3 + 7=	8 + 2=
5 + 6=	6 + 4=	1 + 9=
8 + 3=	2 + 8=	7 + 2=
5 + 6=	2 + 9=	8 + 1=
4 + 6=	6 + 3=	3 + 6=
11 - 1=	7 + 2=	6 + 4=
12 - 2=	1 + 9=	5 + 5=
13 - 3=	3 + 8=	5 + 4=
14 - 4=	2 + 8=	6 + 5=
15 - 5=	4 + 5=	7 + 4=
16 - 6=	4 + 6=	8 + 2=
17 - 7=	7 + 2=	9 + 1=
18 - 8=	9 + 2=	9 + 2=
19 - 9=	8 + 3=	6 + 4=

Ho Math Chess 何数棋谜　妈！我会棋谜式加法啦！

Mom! I Learn Addition Using Math-Chess-Puzzles Connection

Contents include both traditional and Math-Chess-Puzzles combined methods. Extra strength

Student Name _____ Date _____

Matching 2 digits in columns A an B such that their sum is 10.

A	B
1	7
3	9
5	3
7	5
2	6
4	8
8	2
6	4
9	1

A	B
7	2
9	1
3	5
5	8
6	7
8	6
2	3
4	4
1	9

A	B
5	6
6	2
7	5
2	3
3	9
1	4
4	7
8	8
9	1

A	B
2	6
4	4
6	5
7	7
9	3
1	2
3	1
5	8
8	9

Student Name _____ Date _____

Filling each square box by a digit.

8 + 2 = ☐	2 + 8 = ☐
8 + ☐ =10	2 + ☐ = 10
☐ + 2 = 10	☐ + 8 = 10
8 ☐ 2 = 10	2 ☐ 8 = 10
8 ☐ ☐ = 10	2 ☐ ☐ = 10
7 + 3 = ☐	3 + 7 = ☐
7 + ☐ =10	3 + ☐ = 10
☐ + 3 = 10	☐ + 7 = 10
7 ☐ 3 = 10	3 ☐ 7 = 10
7 ☐ ☐ = 10	3 ☐ ☐ = 10

Filling each square box by a digit.

$9 + 1 = \square$	$1 + 9 = \square$
$9 + \square = 10$	$1 + \square = 10$
$\square + 1 = 10$	$\square + 9 = 10$
$9 \,\square\, 1 = 10$	$1 \,\square\, = 10$
$9 \,\square\, \square = 10$	$1 \,\square\, \square$
$6 + 4 = \square$	$4 + 6 = \square$
$6 + \square = 10$	$4 + \square = 10$
$\square + 4 = 10$	$\square + 6 = 10$
$6 \,\square\, 4 = 10$	$4 \,\square\, 6 = 10$
$6 \,\square\, \square = 10$	$4 \,\square\, \square = 10$

each ☐ with a number.

Math and Chess sentence	Number sentence.	Comments
5 + ♟♟♟♟♟	$$5 + 5 = 10$$	
2+ ♟♟♟♟♟♟♟♟		Count right to left to make a ten. (Add the bigger number to the smaller number.)
♟♟♟♟♟♟♟+ 3		Count left to right to make a ten.
1+ ♟♟♟♟♟♟♟♟♟		Count right to left to make a ten.
♟♟♟♟♟♟♟♟+ 2		Count left to right to make a ten.
4 + ♟♟♟♟♟♟		Count right to left to make a ten.
♟♟♟♟♟♟♟+3		
2+ ♟♟♟♟♟♟♟♟		
6 + ♟♟♟♟		
♟♟♟♟♟+5		
4 + ♟♟♟♟♟♟		
6 + ♟♟♟♟		

10's complement

10's complement	10's complement	Make a ten
$10 - 4 = \square$	$10 - 6 = \square$	Four plus six is $\square$
$10 - 1 = \square$	$10 - 9 = \square$	One plus nine is $\square$
$10 - 3 = \square$	$10 - 7 = \square$	Three plus seven is $\square$
$10 - 5 = \square$	$10 - 5 = \square$	Five plus five is $\square$
$10 - 7 = \square$	$10 - 3 = \square$	Seven plus three is $\square$
$10 - 9 = \square$	$10 - 1 = \square$	Nine plus one is $\square$
$10 - 6 = \square$	$10 - 4 = \square$	Six plus four is $\square$
$10 - 2 = \square$	$10 - 8 = \square$	Two plus eight is $\square$
$10 - 1 = \square$	$10 - 9 = \square$	One plus nine is $\square$
$10 - 8 = \square$	$10 - 2 = \square$	Eight plus two is $\square$

♔ = 0, ♙ = 1, ♝ = 3, ♞ = 3, ♖ = 5, ♛ = 9

	♙			
+ 1	+ 9	+ 8	+ 2	+ 7
10		10	10	10
7			♖	♝
+ 3	+ 4	+ 6	+ 5	+ 7
	10	10		
			♙	
+ ♝	+ 8	+ 2	+ 9	+ ♙
10	10	10		10
			♝	
+ 2	+ 8	+ ♝	+ 7	+ 6
10	10	10		10
♖			♝	7
+ 5	+ 6	+ 4	+ 7	+ ♝
	10	10		

Mom! I Learn Addition Using Math-Chess-Puzzles Connection
Contents include both traditional and Math-Chess-Puzzles combined methods. Extra strength
Student Name _____ Date _____

Making ten

※ ♔ = 0, ♙ = 1, ♝ = 3, ♞ = 3, ♖ = 5, ※ ♕ = 9

			♖	♗
+ ♗	+ 4	+ 6	+ 5	+ 7
10	10	10		
			♙	
+ 3	+ 8	+ 2	+ 9	+ 1
10	10	10	10	10
			♞	
+ 2	+ 8	+ 3	+ 7	+ 6
10	10	10	10	10
			♗	
+ ♖	+ 6	+ 4	+ 7	+ ♞
10	10	10		10
	♙			♞
+ 1	+ 9	+ 8	+ 2	+ 7
10		10	10	

Contents include both traditional and Math-Chess-Puzzles combined methods. Extra strength

Student Name _____ Date _____

Associative property 結合律

$(a + b) + c = a + (b + c) = (a + c) + b$

$1 + 2 + 3 = (1 + 2) + 3 = \Box\ +\ 3 = \Box$

$1 + 2 + 3 = 1 + (2 + 3) = 1 + \Box\ =\ \Box$

$1 + 2 + 3 = (1 + 3) + 2 = \Box\ +\ 2 = \Box$

♔ = 0, ♙ = 1, ♗ = 3, ♘ = 3, ♖ = 5, ♕ = 9

·7	·6	4	.♖	·3
2	♗	·4	4	♖
+ ·3	+ ·4	+ ·6	+ ·5	+ ·7

·9	·6	4	♖	♗
2	♖	.♗	·2	·4
+ ·1	+ ·4	+ ·7	+ ·8	+ ·6

Student Name _____ Date _____

Memorization of adding to 10

Fill in □ with a number.

❋ ♔ = 0, Ψ ♙ = 1, ✗ ♟ = 3, ✚ ♞ = 3, ✚ ♜ = 5, ❋ ♛ = 9

Ψ　✚　✚　❋　✗　❋

1		2		✚♞		4		♜	
□	♛	□		□	7	□		□	6
10	□	10	8	10	□	10	6	10	□
	10		□		10		□		10
4		✚♞	10	2		♗	10	4	
□	7	□		□	♛	□		□	✚♞
10	□	10	3	10	□	10	8	10	□
	10		□		10		□		10
♗		2	10	✚♞		4	10	♜	
□	9	□		□	7	□		□	6
10	□	10	4	10	□	10	6	10	□
	10		□		10		□		10
4		✚♞	10	2		♗	10	6	
□	7	□		□	♛	□		□	✚♞
10	□	10	5	10	□	10	8	10	□
	10		□		10		□		10
			10				10		

Ho Math Chess 何数棋谜 妈！我会棋谜式加法啦！
Mom! I Learn Addition Using Math-Chess-Puzzles Connection
Contents include both traditional and Math-Chess-Puzzles combined methods. Extra strength

Student Name _____ Date _____

Memorization of adding to 10

Fill in □ with a number.

(star)(king) = 0, (pawn) = 1, (skull)(bishop) = 3, (knight) = 3, (rook) = 5, (star)(queen) = 9

	2		5		6		2		7
1	□	2	□	(knight)	□	4	□	(rook)	□
□	10	□	10	□	10	□	10	□	10
10		10		10		10		10	
	6	(queen)		(knight)		4		8	
4	□	(knight)	□	2	□	1	□	6	□
□	10	□	10	□	10	□	10	□	10
10		10		10		10		10	
(queen)		6		2		4		6	
7	□	2	□	(bishop)	□	4	□	(rook)	□
□	10	□	10	□	10	□	10	□	10
10		10		10		10		10	
	6		4		8		3		7
4	□	3	□	2	□	1	□	6	□
□	10	□	10	□	10	□	10	□	10
10		10		10		10		10	
(bishop)		7		6		4		(queen)	
	□		□		□		□		□
	10		10		10		10		10

Grouping 10

Fill in the following the ☐ with a number.

✳ ♔ = 0, ♕ ♙ = 1, ✖ ♗ = 3, ✚ ♘ = 3, ✚ ♖ = 5, ✳ ♛ = 9

$$6 + 4 + 2 + ♖ + 3 + ♗ + 7 = \boxed{}$$

$$3 + 2 + 7 + 8 + 2 = \boxed{}$$

$$♙ + 2 + 6 + 9 + 4 = \boxed{}$$

$$8 + 4 + ♖ + 6 + 5 + 2 = \boxed{}$$

$$♙ + 6 + 9 + 4 + 9 = \boxed{}$$

$$2 + \boxed{} + 8 + 7 + 5 = \boxed{}$$

$$9 + 6 + 3 + 7 + 4 + ♙ + 5 + ♖ = \boxed{}$$

$$2 + \boxed{} + 6 + 7 + 4 + 8 + 1 + 9 + 3 = \boxed{}$$

Student Name _____ Date _____

Addition strategy: 9 plus 1 (9 + d = 9 + 1 + (d−1))　(9 加 1, 9 + d= 9 + 1 + (d−1))

$$9 + 6 = 9 + 1 + 5 = 10 + 5 = 15$$

❋ ♔ = 0, ♕ ♙ = 1, ✖ ♗ = 3, ✚ ♞ = 3, ✚ ♖ = 5, ❋ ♛ = 9

The pre-requisite to use the above strategy is that the students know how to partition a number into a sum of 1 + a smaller number such as 4 = 1 + 3, 5 = 1+ 4 etc. I have taught some students who got confused on this strategy, one of the reasons is they do not understand 9 can be 1 + 8.

4 = 1 + ____
6 = 1 + ____
5 = 1 + ____
7 = 1 + ____
8 = 1 + ____
9 = 1 + ____
3 = 1 + ____
2 = 1 + ____
4 = ____ + 1
5 = ____ + 1
7 = ____ + 1
6 = ____ + 1
8 = ____ + 1
9 = ____ + 1
4 = ____ + 3
6 = ____ + 5
5 = ____ + 4

If 10 + 1 = 11, then 9 + 2 = 9 + 1 + ___ = 10 + ___ = ____

If 10 + 2 = 12, then 9 + 3 = 9 + 1 + ___ = 10 + ___ = ____

If 10 + 3 = 13, then 9 + 4 = 9 + 1 + ___ = 10 + ___ = ____

If 10 + 4 = 14, then 9 + 5 = 9 + 1 + ___ = 10 + ___ = ____

If 10 + 5 = 15, then 9 + 6 = 9 + 1 + ___ = 10 + ___ = ____

If 10 + 6 = ___, then 9 + 7 = 9 + 1 + ___ = 10 + ___ = ____

If 10 + 7 = ___, then 9 + 8 = 9 + 1 + ___ = 10 + ___ = ____

If 10 + 8 = ___, then 9 + 9 = double 9 = = ____

Student Name _____ Date _____

9	6	9	2	9
+ 5	+ 9	+ 2	+ 9	+ 3
5	9	9	8	9
+ 9	+ 4	+ 7	+ 9	+ 5
9	3	9	4	5
+ 6	+ 9	+ 6	+ 9	+ 9
9	6	7	9	9
+ 8	+ 9	+ 9	+ 8	+ 8
8	7	6	9	3
+ 9	+ 9	+ 9	+ 5	+ 9

Student Name _____ Date _____

♔ = 0, ♕ ♙ = 1, ♗ = 3, ♘ = 3, ♖ = 5, ♛ = 9

3	10	9	4	10
+ ♛	+ 4	+ 4	+ 9	+ 5
9	♖	10	♕	6
+ ♖	+ 9	+ 6	+ 6	+ 9
10	♕	7	10	9
+ 7	+ 7	+ 9	+ 8	+ 8
8	10	♕	10	♕
+ 9	+ 9	+ 9	+ 7	+ 7

Student Name _____ Date _____

Adding 9 by relating

Adding 9 +d by splitting the other number d to be 1 and d − 1

✳ ♔ = 0, ♕ ♙ = 1, ✳ ♝ = 3, ✿ ♘ = 3, ✛ ♖ = 5, ✳ ♛ = 9

Adding 9 by relating

Adding 9 +d by splitting the other number d to be 1 and d − 1

(♚ = 0, ♟ = 1, ♝ = 3, ♞ = 3, ♜ = 5, ♛ = 9)

Adding 9 by relating

Adding 9 +d by splitting the other number d to be 1 and d − 1

✳ ♔ = 0, ♙ ♙ = 1, ✕ ♗ = 3, ✜ ♘ = 3, ✛ ♖ = 5, ✳ ♕ = 9

Adding 9 by relating

Adding 9 +d by splitting the other number d to be 1 and d − 1

(♔ = 0, ♙ = 1, ♗ = 3, ♘ = 3, ♖ = 5, ♕ = 9)

Using 9 plus technique

3	229	345	564
2	765	**999**	989
1	548	888	899
	a	b	c

The original square is at b2 = ☐.

0708090204040503
+ 9909090909090909

b2 + ✛ =

☐☐☐
+ ○○○

☐ + ✕ =

☐☐☐
+ ○○○

b2 + ✛ =

☐☐☐
+ ○○○

☐ + ✛ = 1332

☐☐☐
+ ○○○

1764

b2 + ✕ =

☐☐☐
+ ○○○

☐ + ✛ =

☐☐☐
+ ○○○

b2 + ✕ =

☐☐☐
+ ○○○

☐ + ✕ =

☐☐☐
+ ○○○

Contents include both traditional and Math-Chess-Puzzles combined methods. Extra strength

Student Name _____ Date _____

9 plus
Adding 9 +d by splitting the other number d to be 1 and d − 1

5		9		9	
4	8	6	1	7	6
3		3	9	2	
2	5	8	4	5	7
1		9		9	
	a	b	c	d	e

You are at square c3

= ▢ .

Student Name _____　Date _____

9 plus
Adding 9 +d by splitting the other number d to be 1 and d − 1

5		9		9	
4	8	6	1	7	6
3		3	9	2	
2	5	8	4	5	7
1		9		9	
	a	b	c	d	e

You are at square c3

= ☐ .

Adding 9 +d by splitting the other number d to be 1 and d − 1

You are a chess piece located at a square indicated by a shaded square.

Student Name _____ Date _____

Adding 9 +d by splitting the other number d to be 1 and d − 1

1	1	2
3	9	1
3	2	2

You are a chess piece located at a square indicated by a shaded square.

○ = +

Student Name _____ Date _____

Adding 9 +d by splitting the other number d to be 1 and d − 1

You are a chess piece located at a square indicated by a shaded square.

$\bigcirc = +$

Student Name _____ Date _____

Adding 9 +d by splitting the other number d to be 1 and d − 1

2	2	3
1	9	3
2	1	1

You are a chess piece located at a square indicated by a shaded square.

○ = +

Adding 9 +d by splitting the other number d to be 1 and d − 1

You are a chess piece located at a square indicated by a shaded square.

○ = +

Adding 9 +d by splitting the other number d to be 1 and d − 1

8	2	9
5	**9**	3
7	4	6

You are a chess piece located at a square indicated by a shaded square.

○ = +

Adding 9 +d by splitting the other number d to be 1 and d − 1
Only transfer each number to a correct position. Do not the transfer the 3 by 3 square.

You are a chess piece located at a square indicated by a shaded square.

○ = +

Adding 9 +d by splitting the other number d to be 1 and d − 1
Only transfer each number to a correct position. Do not the transfer the 3 by 3 square.

Student Name _____　Date _____

You are a chess piece located at a square indicated by a shaded square.

$\bigcirc = +$

Student Name _____ Date _____

Adding 9 +d by splitting the other number d to be 1 and d − 1

Only transfer each number to a correct position. Do not the transfer the 3 by 3 square.

You are a chess piece located at a square indicated by a shaded square.

○ = +

Adding 9 +d by splitting the other number d to be 1 and d − 1

Only transfer each number to a correct position. Do not the transfer the 3 by 3 square.

You are a chess piece located at a square indicated by a shaded square.

$\bigcirc = +$

Using doubles, 9 + d to add

$9 + d = 9 + 1 + (d-1)$, $9 + 6 = 9 + \mathbf{1} + \mathbf{5} = 15$

6 + 7 = 13   7 + 8 = 15   5 + 6 = 11

7 + 8 = 15   6 + 8 = 14   6 + 9 = 15

3 + 5 = 8   7 + 9 = 16   8 + 9 = 17

5 + 6 = 11   5 + 6 = 11   5 + 7 = 12

7 + 9 = 16   6 + 8 = 14   5 + 6 = 11

5 + 7 = 12   8 + 9 = 17   7 + 8 = 15

Using doubles or 9 + d to add

(♚ = 0, ♙ = 1, ♗ = 3, ♘ = 3, ♖ = 5, ♕= 9)

♕ ♜ ♜ ✷ ✕ ✷

2	2	♗	♗	4
+ 2	+ ♞	+ ♗	+ 4	+ 4
3	♞	4	4	4
+ 2	+ 3	+ 3	+ 4	+ ♖
5	♖	6	6	7
+ 5	+ 6	+ 6	+ 7	+ 7
6	6	7	7	8
+ 5	+ 6	+ 6	+ 7	+ 7
8	8	♕	7	♕
+ 8	+ 9	+ ♕	+ 8	+ 8

Using doubles or 9 + d to add

(♚ = 0, ♙ = 1, ♗ = 3, ♞ = 3, ♖ = 5, ♛ = 9)

2	3	4	5	6
+ 2	+ 3	+ 4	+ 5	+ 6
7	8	9	♞	4
+ 7	+ 8	+ 9	+ ♞	+ 4
1 3	1 3	♖	6	6
+ ♞	+ 4	+ ♖	+ 5	+ 6
6	7	7	7	8
+ 7	+ 6	+ 7	+ 8	+ 7
8	8	♛	9	6
+ 8	+ ♛	+ 8	+ 9	+ 5

Using doubles or 9 + d to add

(♔ = 0, ♙ = 1, ♗ = 3, ♘ = 3, ♖ = 5, ♕ = 9)

2	♗	4	♖	9
+ 2	+ ♗	+ 4	+ 5	+ 9
7	8	♕	2	2
+ 7	+ 8	+ ♕	+ 2	+ ♗
8	♖	6	5	4
+ 8	+ 5	+ 5	+ 6	+ 3
♕	7	8	♕	8
+ 9	+ 6	+ 7	+ 8	+ 9
9	♖	7	6	6
+ 8	+ 6	+ 6	+ 7	+ ♖

Student Name _____ Date _____

Using doubles or 9 + d to add

(♔ = 0, ♙ = 1, ♗ = 3, ♘ = 3, ♖ = 5, ♕ = 9)

7	8	8	♖	7
+ 6	+ 9	+ 7	+ 8	+ 5
7	8	8	6	5
+ 8	+ 9	+ 7	+ ♖	+ 7
♕	5	6	7	7
+ 8	+ 6	+ ♖	+ 8	+ 5

Ho Math Chess 何数棋谜   妈！我会棋谜式加法啦！
Mom! I Learn Addition Using Math-Chess-Puzzles Connection

Contents include both traditional and Math-Chess-Puzzles combined methods. Extra strength

©2008 – 2018 Frank Ho, Amanda Ho   All rights reserved. www.homathchess.com

Student Name _____ Date _____

Using doubles or 9 + d to add

($\text{♔} = 0$, $\text{♙} = 1$, $\text{♗} = 3$, $\text{♘} = 3$, $\text{♖} = 5$, $\text{♕} = 9$)

9	7	8	♕	8
+ 8	+ 6	+ 7	+ 8	+ 6
♕	♖	7	6	6
+ 8	+ 6	+ 6	+ 7	+ 5

Ho Math Chess 何数棋谜　妈！我会棋谜式加法啦！

Mom! I Learn Addition Using Math-Chess-Puzzles Connection

Contents include both traditional and Math-Chess-Puzzles combined methods. Extra strength

Student Name _____　　Date _____

Using doubles or 9 + d to add

(♔ = 0, ♙ = 1, ♗ = 3, ♘ = 3, ♖ = 5, ♕ = 9)

♕ ♙ ♙ ♖ ♘ ♖

7	8	8	♖	7
+ 6	+ 9	+ 7	+ 8	+ 5
7	8	8	6	5
+ 8	+ 9	+ 7	+ ♖	+ 7
♕	♖	6	7	7
+ 8	+ 6	+ 5	+ 8	+ 5
9	7	8	♕	8
+ 8	+ 6	+ 7	+ 8	+ 6

Adding 5 by doubling and 10's complement

		1		5	
4	6(5+1)	6(5+1)	4	7 (5+2)	3
3		3	5	6(5+1)	
2	8(5+3)	4	5	8(5+3)	7(5+2)
1		9		2	
	a	b	c	d	e

You are at square c3

= ☐ .

☐ + ⬚ = __

☐ + ⬚ = __

☐ + ⬚ = __

☐ + ⬚ = __

☐ + ⬚ = __

☐ + ⬚ = __

☐ + ⬚ = __

☐ + ⬚ = __

Student Name _____ Date _____

Adding 5 by doubling and 10's complement

5		9		9	
4	8(5+3)	6	1	7(5+2)	6
3		3	⬖	2	
2	5	8(5+3)	4	5	7(5+2)
1		9		9	
	a	b	c	d	e

You are at square c3

= ☐ .

Adding 6 by doubling and 10's complement

5		9		9	
4	8(6+2)	6	1	7(6+1)	6
3		3	6	2	
2	5	8(6+2)	4	5	7(6+1)
1		9		9	
	a	b	c	d	e

You are at square c3

= [   ].

Adding 6 by doubling and 10's complement

	a	b	c	d	e
5		9		9	
4	8(6+2)	6	1	7	6
3		3	6	2	
2	5	8(6+2)	4	5	7
1		9		9	
	a	b	c	d	e

You are at square c3

$= \boxed{}$.

Student Name _____ Date _____

Adding 7 by doubling and 10's complement

5		9		9	
4	8(7+1)	6	1	7	6
3		3	7	2	
2	5	8(7+1)	4	5	7
1		9		9	
	a	b	c	d	e

You are at square c3

= ⬜ .

Mom! I Learn Addition Using Math-Chess-Puzzles Connection
Contents include both traditional and Math-Chess-Puzzles combined methods. Extra strength

Student Name _____ Date _____

Adding 7 by doubling and 10's complement

5		9		9	
4	8(7+1)	6	1	7	6
3		3	7	2	
2	5	8(7+1)	4	5	7
1		9		9	
	a	b	c	d	e

You are at square c3

= ☐ .

Student Name _____ Date _____

Adding 8 by doubling and 10's complement

5		9		9	
4	8	6	1	7	6
3		3	8	2	
2	5	8	4	5	7
1		9		9	
	a	b	c	d	e

You are at square c3

= ☐ .

Adding 8 by doubling

5		9		9	
4	8	6	1	7	6
3		3	8	2	
2	5	8	4	5	7
1		9		9	
	a	b	c	d	e

You are at square c3

= ☐ .

Making 10 or adding around 10 or doubling

7 + 5 = 10

9 + 3 = 12

5 + 9 = 14

5 + 8 = 13

7 + 4 = 11

6 + 8 = 14

6 + 5 = 11

8 + 6 = 14

9 + 4 = 13

9 + 4 = 13

6 + 9 = 15

7 + 4 = 11

7 + 8 = 15

6 + 7 = 13

8 + 5 = 13

Student Name _____ Date _____

Adding up to 10 or doubling

Use the following strategies:

9 + number to split the number to be 1 + (number −1). 9 + 5 = 9 + 1 + 4
Use doubling whenever you can. 7 + 8 = 7 + 7 + 1 = 14 + 1
Use adding up to 10. 7 + 4 = 7 + 3 + 1 = 10 + 1 = 11

1	$7 + 4 = 7 + (\ ) + 1 = (\ ) + 1 = (\ )$
2	$7 + 5 = 5 + (\ ) + 2 = (\ ) + 2 = (\ )$
3	$7 + 6 = 6 + (\ ) + 1 = (\ ) + 1 = (\ )$
4	$7 + 8 = 7 + 7 + (\ ) = (\ ) + 1 = (\ )15$
5	$7 + 9 = 6 + 1 + (\ ) = (\ ) + 6 = (\ )$
6	$6 + 5 = 1 + (\ ) + 5 = (\ ) + 1 = (\ )$
7	$6 + 6 = (\ )$
8	$6 + 7 = 6 + 6 + (\ ) = 12 + (\ ) = (\ )$
9	$6 + 8 = 6 + 6 + (\ ) = 12 + (\ ) = (\ )$
10	$6 + 9 = 9 + 1 + (\ ) = (\ ) + 5 = (\ )$
11	$7 + 8 = 7 + 7 + (\ ) = 1 + (\ ) = (\ )$
12	$5 + 6 = 5 + 5 + (\ ) = (\ ) + 1 = (\ )$
13	$5 + 7 = 5 + 5 + (\ ) = (\ ) + 2 = (\ )$ 5+5+2=10+2=12

Mixed doubling, adding to 10, adding 1, 2, 3

Evaluate the following results.

$$\overset{\displaystyle 12}{\overset{\displaystyle \frown}{\underset{\displaystyle 6+8+2+6}{\overset{\displaystyle 10}{\frown}}}} = 22$$

$7 + 8 + 3 + 8$

$4 + 8 + 2 + 4$

$5 + 6 + 5 + 4$

$9 + 3 + 3 + 1$

$2 + 5 + 5 + 8$

$4 + 2 + 4 + 8$

$7 + 1 + 7 + 9$

$9 + 2 + 9 = 8$

$4 + 8 + 8 + 6$

Student Name _____ Date _____

Mixed doubling and adding to 10

9 + 1 + 5 + 5
6 + 2 + 8 + 6
7 + 5 + 3 + 5
2 + 7 + 8 + 7
8 + 8 + 4 + 6
7 + 3 + 7 + 7
5 + 5 + 5 + 5
2 + 8 + 6 + 6
8 + 8 + 1 + 9
7 + 3 + 8 + 8

Student Name ______________________________ Date ____________________

8 + 8 + 8 + 2
6 + 6 + 7 + 3
7 + 7 + 7 + 3
5 + 5 + 6 + 5
6 + 6 + 4 + 6
4 +8 + 2 + 4
9 + 6 + 1 + 6
9 + 9 + 9 + 1
1 + 8 + 1+ 2
7 + 3 + 3 + 3

Student Name _____ Date _____

Mixed doubling, adding to 10, adding 1, 2, 3

Evaluate the following results.

$$6 + 8 + 2 + 6 + 2 = 24$$

$7 + 8 + 3 + 8 + 1$

$4 + 8 + 2 + 4 + 2$

$5 + 6 + 5 + 4 + 3$

$9 + 3 + 3 + 1 + 2$

$2 + 5 + 5 + 8 + 1$

$4 + 2 + 4 + 8 + 3$

$7 + 1 + 7 + 9 + 3$

$9 + 2 + 9 + 8 + 2$

$4 + 8 + 8 + 6 + 1$

27, 20, 23, 18, 21, 21, 27, 30, 27

Multi-direction addition of doubling and adding to 10

$$6\ (5+1)$$

$$+\ \text{♜}$$

$$6 + \text{♜} = 5 + 5 + \square = \square = 6 + \square + \text{♟}$$

$$\llcorner\ 10\ \lrcorner \qquad\qquad\qquad \llcorner\ 10\ \lrcorner$$

$$5$$

$$+\ \text{♜}$$

$$7 + 6 = 6 + 6 + \square = \square 13 = 3 + \square 7 + \text{♗}$$

$$\llcorner\ 12\ \lrcorner \qquad\qquad\qquad \llcorner\ 10\ \lrcorner$$

Student Name ______________________ Date ______________

  10
\+ ♗
____

8 + ♜ = 5 + 5 + = □ = 8 + □ + ♗
└ 10 ┘ └ 10 ┘

  6
\+ ♜
____

9 + 2 = 9 + 1 + □ = □ = 8 + □ + ♟
└ 10 ┘ └ 10 ┘

If-Then for plus (do the left + first then the right +) review

$$10 \quad - \quad ♙ \quad = \quad 9$$
$$+ \quad 7 \quad + \quad ♔ \quad = \quad 7 \qquad +$$
_____

If 10 + 7 = ☐, then ♛ + 7 must be ☐.

If 8 + 8 = ☐, then ♛ + 7 must be ☐.

$$10 \quad - \quad ♙ \quad = \quad 9$$
$$+ \quad 6 \quad + \quad ♔ \quad = \quad 6 \qquad +$$
_____

If 10 + 6 = ☐, then ♛ + 6 must be ☐.

If 6 + 6 = ☐, then ♛ + 6 must be ☐.

$$7 \quad + \quad ♙ \quad = \quad 8$$

$$+ \quad 7 \quad + \quad ♚ \quad = \quad 7 \quad +$$

_____   _____

If 7 + 7 = ☐ , then 7 + 8 must be ☐ .

If 8 + 8 = ☐ , then 8 + 7 must be ☐ .

$$10 \quad - \quad ♙ \quad = \quad 9$$

$$+ \quad 5 \quad + \quad ♚ \quad = \quad 5 \quad +$$

_____   _____

If 10 + ♖ = ☐ , then ♕ + ♖ must be ☐ .

If ♖ + 10 = ☐ , then ♖ + ♕ must be ☐ .

Student Name _____ Date _____

$$6 \quad + 2 \quad = \quad 8$$

$$+ \quad 6 \quad + \; ♔ \quad = \quad 6 \quad +$$

_____　　　_____

If 6 + 6 = ☐, then 8 + 6 must be ☐.

If 6 + 6 = ☐, then 6 + 8 must be ☐.

$$10 \quad - \; ♙ \quad = \quad 9$$

$$+ \quad 4 \quad + \; ♔ \quad = \quad 4 \quad +$$

_____　　　_____

If 10 + 4 = ☐, then ♕ + 4 must be ☐.

If 4 + 10 = ☐, then 4 + ♕ must be ☐.

♖ + ♗ = **8**

+   ♖ + ♔ = **5**   +

_____

If ♖ + ♖ = , then 8 + ♖ must be .

If ♖ + ♖ = , then ♖ + 8 must be .

6   + ♙ = **7**

+   6   + ♚ = **6**   +

_____

If 6 + 6 = □ , then 7 +

6 must be □ .

If 6 + 6 = □ , then 6 + 7 must be □ .

10　　–　♙　=　**9**

+　　　♝　+　♔　=　**3**　　+

_____　　　　_____

If 10 + ♝ = **13**, then ♛ + ♝ must be .

If ♝ + 10 = , then ♝ + ♛ must be .

4　　+ 4　　=　**8**

+　　　4　　+ ♔　=　**4**　　+

_____　　　　_____

If 4 + 4 = ☐, then 8 + 4 must be ☐.

If 4 + 4 = ☐, then 4 + 8 must be ☐.

♖ + 2 = 7

+　　♖ + ♔ = 5　　+

_____　　_____

If ♖ + ♖ = □, then 7 + ♖ must be □.

If ♖ + ♖ = □, then ♖ + 7 must be □.

9　+ ♔ = 9

+　　♙ + ♙ = 2　　+

_____　　_____

If ♕ + ♙ = □10, then ♕ + 2 must be □11.

If ♙ + ♕ = □10, then 2 + ♕ must be □11.

8　+　♔　=　**8**

+　　2　+　♙　=　**3**　+

_____　　_____

If 8 + 2 = ☐, then 8 + ♗ must be ☐.

If 2 + 8 = ☐, then ♗ + 8 must be ☐.

7　+　♔　=　**7**

+　　3　+　♙　=　**4**　+

_____　　_____

If 7 + ♗ = ☐, then 7 + 4 must be ☐.

If ♗ + 7 = ☐, then 4 + 7 must be ☐.

Student Name ______________________________ Date ______________

♖ + ♙ = 6

\+ ♖ + ♔ = 5 +

__________ __________

If ♖ + ♖ = □, then 6 + ♖ must be □.

If ♖ + ♖ = □, then ♖ + 6 must be □.

Student Name _____ Date _____

Assessment of 9 + d, doubling, adding to 10

	D	T	123	DT
D	$\bigcirc + \bigcirc = 10$	$\bigcirc + \bigcirc + 6 + 4 = 18$ The following has 5 answers $\bigcirc + \bigcirc + 7 + \triangle = 18$ $\bigcirc + \bigcirc + 7 + \triangle = 18$ $\bigcirc + \bigcirc + 7 + \triangle = 18$ $\bigcirc + \bigcirc + 7 + \triangle = 18$ $\bigcirc + \bigcirc + 7 + \triangle = 18$	$\bigcirc + \bigcirc + 3 = 11$	
T		$6 + 4 + 7 + \triangle = 20$	$6 + 4 + \triangle = 13$ $\triangle + 7 + 3 = 12$	
123			$2 + 4 + \bigcirc = 9$	$\triangle + \triangle + 7 + 3 = 26$

Part 3: More practices of chess and math integrated problems

Students have finished all addition strategies, so let's have some fun to review them. If students need more practices, then this part offers more practices.

Additions to 10

5	?	4	?	?	?
4	2	9	3	2	5
3	?	6	0	4	?
2	?	8	5	7	?
1	?	?	?	7	?
	a	b	c	d	e

You are at square c3

= ☐ .

Student Name _____ Date _____

Adding up to 10

5	?	?	?	?	?
4	2	9	3	2	5
3	?	6	0	4	?
2	?	8	5	7	?
1	?	?	?	7	?
	a	b	c	d	e

You are at square c3

$= \boxed{}$.

Ho Math Chess 何数棋谜　妈！我会棋谜式加法啦！
Mom! I Learn Addition Using Math-Chess-Puzzles Connection
Contents include both traditional and Math-Chess-Puzzles combined methods. Extra strength

©2008 − 2018 Frank Ho, Amanda Ho　All rights reserved. www.homathchess.com

Student Name _____ Date _____

Making 10 or adding around 10

Only transfer each number to a correct position. Do not the transfer the 3 by 3 square.

2	2	1
2	**8**	1
1	3	3

You are a chess piece located at a square indicated by a shaded square.

○ = +

Making 10 or adding around 10

Only transfer each number to a correct position. Do not the transfer the 3 by 3 square.

1	2	2
3	8	2
3	1	1

You are a chess piece located at a square indicated by a shaded square.

○ = +

Student Name _____ Date _____

Making 10 or adding around 10
Only transfer each number to a correct position. Do not the transfer the 3 by 3 square.

You are a chess piece located at a square indicated by a shaded square.

◯ = +

Making 10 or adding around 10

Only transfer each number to a correct position. Do not the transfer the 3 by 3 square.

You are a chess piece located at a square indicated by a shaded square.

$$\bigcirc = +$$

Making 10 or adding around 10

Only transfer each number to a correct position. Do not the transfer the 3 by 3 square.

4	3	2
3	7	2
3	4	4

You are a chess piece located at a square indicated by a shaded square.

○ = +

Mom! I Learn Addition Using Math-Chess-Puzzles Connection

Contents include both traditional and Math-Chess-Puzzles combined methods. Extra strength

Student Name _____ Date _____

Making 10 or adding around 10

Only transfer each number to a correct position. Do not the transfer the 3 by 3 square.

3	3	4
4	7	3
4	2	2

You are a chess piece located at a square indicated by a shaded square.

○ = +

Making 10 or adding around 10
Only transfer each number to a correct position. Do not the transfer the 3 by 3 square.

You are a chess piece located at a square indicated by a shaded square.

○ = +

Student Name _____ Date _____

Making 10 or adding around 10

Only transfer each number to a correct position. Do not the transfer the 3 by 3 square.

You are a chess piece located at a square indicated by a shaded square.

$$\bigcirc = +$$

Student Name _____
Date _____

Making 10 or adding around 10 or doubling
Only transfer each number to a correct position. Do not the transfer the 3 by 3 square.

3	4	3
4	6	3
3	5	4

You are a chess piece located at a square indicated by a shaded square.

○ = +

Making 10 or adding around 10 or doubling
Only transfer each number to a correct position. Do not the transfer the 3 by 3 square.

You are a chess piece located at a square indicated by a shaded square.

○ = +

Making 10 or adding around 10 or doubling

Only transfer each number to a correct position. Do not the transfer the 3 by 3 square.

You are a chess piece located at a square indicated by a shaded square.

$\bigcirc = +$

Making 10 or adding around 10 or doubling
Only transfer each number to a correct position. Do not the transfer the 3 by 3 square.

3	3	4
4	**6**	5
3	4	3

You are a chess piece located at a square indicated by a shaded square.

○ = +

Making 10 or adding around 10 or doubling

4 5 4 5 5 4 6 6 5	You are a chess piece located at a square indicated by a shaded square. ○ = +

Making 10 or adding around 10 or doubling

Only transfer each number to a correct position. Do not the transfer the 3 by 3 square.

You are a chess piece located at a square indicated by a shaded square.

○ = +

Student Name _____ Date _____

Making 10 or adding around 10 or doubling
Only transfer each number to a correct position. Do not the transfer the 3 by 3 square.

You are a chess piece located at a square indicated by a shaded square.

○ = +

Ho Math Chess 何数棋谜　妈！我会棋谜式加法啦！

Mom! I Learn Addition Using Math-Chess-Puzzles Connection

Contents include both traditional and Math-Chess-Puzzles combined methods. Extra strength

©2008 — 2018 Frank Ho, Amanda Ho　　All rights reserved. www.homathchess.com

Student Name _____ Date _____

Making 10 or adding around 10 or doubling

Only transfer each number to a correct position. Do not the transfer the 3 by 3 square.

You are a chess piece located at a square indicated by a shaded square.

$\bigcirc = +$

Student Name _____ Date _____

Making 10 or adding around 10 or doubling
Only transfer each number to a correct position. Do not the transfer the 3 by 3 square.

You are a chess piece located at a square indicated by a shaded square.

○ = +

Student Name _____ Date _____

Making 10 or adding around 10 or doubling

Only transfer each number to a correct position. Do not the transfer the 3 by 3 square.

You are a chess piece located at a square indicated by a shaded square.

○ = +

Making 10 or adding around 10 or doubling

7	6	7
6	**4**	5
6	7	5

You are a chess piece located at a square indicated by a shaded square.

○ = +

Student Name _____ Date _____

Making 10 or adding around 10 or doubling
Only transfer each number to a correct position. Do not the transfer the 3 by 3 square.

You are a chess piece located at a square indicated by a shaded square.

○ = +

Student Name _____ Date _____

Making 10 or adding around 10 or doubling

Only transfer each number to a correct position. Do not the transfer the 3 by 3 square.

You are a chess piece located at a square indicated by a shaded square.

$$\bigcirc = +$$

Student Name _____ Date _____

Making 10 or adding around 10 or doubling

Only transfer each number to a correct position. Do not the transfer the 3 by 3 square.

You are a chess piece located at a square indicated by a shaded square.

○ = +

Ho Math Chess 何数棋谜   妈！我会棋谜式加法啦！
Mom! I Learn Addition Using Math-Chess-Puzzles Connection
Contents include both traditional and Math-Chess-Puzzles combined methods. Extra strength

©2008 － 2018 Frank Ho, Amanda Ho   All rights reserved. www.homathchess.com

Student Name _____ Date _____

Making 10 or adding around 10

7	7	8
6	**3**	6
6	8	7

You are a chess piece located at a square indicated by a shaded square.

$\bigcirc = +$

Ho Math Chess 何数棋谜　妈！我会棋谜式加法啦！
Mom! I Learn Addition Using Math-Chess-Puzzles Connection
Contents include both traditional and Math-Chess-Puzzles combined methods. Extra strength

©2008 – 2018 Frank Ho, Amanda Ho   All rights reserved. www.homathchess.com

Student Name _____ Date _____

Making 10 or adding around 10

Only transfer each number to a correct position. Do not the transfer the 3 by 3 square.

You are a chess piece located at a square indicated by a shaded square.

○ = +

Student Name _____　Date _____

Making 10 or adding around 10
Only transfer each number to a correct position. Do not the transfer the 3 by 3 square.

7	8	6
6	3	6
8	7	7

You are a chess piece located at a square indicated by a shaded square.

○ = +

Student Name _____ Date _____

Making 10 or adding around 10
Only transfer each number to a correct position. Do not the transfer the 3 by 3 square.

You are a chess piece located at a square indicated by a shaded square.

$\bigcirc = +$

Student Name _____ Date _____

Making 10 or adding around 10

7	8	9
8	**2**	7
8	9	7

You are a chess piece located at a square indicated by a shaded square.

○ = +

Student Name _____ Date _____

Making 10 or adding around 10

Only transfer each number to a correct position. Do not the transfer the 3 by 3 square.

You are a chess piece located at a square indicated by a shaded square.

○ = +

Student Name _____ Date _____

Making 10 or adding around 10

Only transfer each number to a correct position. Do not the transfer the 3 by 3 square.

	You are a chess piece located at a square indicated by a shaded square. ◯ = +

11 11 11 10
10 10 10 9
10 10 10 9
9 9 9 9

Student Name _____ Date _____

Making 10 or adding around 10

Only transfer each number to a correct position. Do not the transfer the 3 by 3 square.

9	7	7
8	2	9
7	8	8

You are a chess piece located at a square indicated by a shaded square.

○ = +

Student Name _____ Date _____

Making 10 or adding around 10

7	9	8
8	1	7
8	9	7

You are a chess piece located at a square indicated by a shaded square.

○ = +

Making 10 or adding around 10

Only transfer each number to a correct position. Do not the transfer the 3 by 3 square.

You are a chess piece located at a square indicated by a shaded square.

○ = +

Mom! I Learn Addition Using Math-Chess-Puzzles Connection

Contents include both traditional and Math-Chess-Puzzles combined methods. Extra strength

Student Name _____ Date _____

Making 10 or adding around 10

Only transfer each number to a correct position. Do not the transfer the 3 by 3 square.

You are a chess piece located at a square indicated by a shaded square.

$$\bigcirc = +$$

Student Name _____ Date _____

Making 10 or adding around 10
Only transfer each number to a correct position. Do not the transfer the 3 by 3 square.

You are a chess piece located at a square indicated by a shaded square.

Knight moves to make 10

For those who know how to move knights.

3	1		
2			
1			
	a	b	c

Step 1
Start at a3 and write the number 1.
Step 2
Make a knight move in clockwise and write a number in the empty square such that the value of a3 + the value of the next move = 10.
Step 3
Continue the knight moves such that the current value plus the value of the next move is 10 until all knight moves are completed.

3

To solve the problem, first you position yourself at the square of a3 which is 1 for this problem.

☐ ☐ ☐ = ☐

☐

☐

You move in the dirction of the knight's move ¬. The answer is 1 + 9 = 10.

The top to down answer for L is also 1 + 9 = 10.

‖

☐

a

Student Name _____　Date _____

Knight moves to make 10

				Step 1
3	8	2	8	Start at a2 and write the number 2.
2	2		2	Step 2
1	8	2	8	Make a knight move in clockwise and write a number in the empty square such that the value of a2 + the value of the next move = 10.
	a	**b**	**c**	Step 3

Step 1
Start at a2 and write the number 2.
Step 2
Make a knight move in clockwise and write a number in the empty square such that the value of a2 + the value of the next move = 10.
Step 3
Continue the knight moves such that the current value plus the value of the next move is 10 until all knight moves are completed.

2

a

Mom! I Learn Addition Using Math-Chess-Puzzles Connection

Contents include both traditional and Math-Chess-Puzzles combined methods. Extra strength

Student Name _____ Date _____

Knight moves to make 10

Step 1
Start at a1 and write the number 3.
Step 2
Make a knight move in clockwise and write a number in the empty square such that the value of a1 + the value of the next move = 10.
Step 3
Continue the knight moves such that the current value plus the value of the next move is 10 until all knight moves are completed.

1

a

Knight moves to make 10

3	6	4	6	Step 1
2	4		4	Start at b1 and write the number 4.
1	6	4	6	Step 2
	a	**b**	**c**	

Step 1
Start at b1 and write the number 4.
Step 2
Make a knight move in clockwise and write a number in the empty square such that the value of b1 + the value of the next move = 10.
Step 3
Continue the knight moves such that the current value plus the value of the next move is 10 until all knight moves are completed.

□ = □□□ ■ □□□ = □

1

b

Knight moves to make 10

3	5	5	5
2	5		5
1	5	5	5
	a	**b**	**c**

Step 1
Start at b3 and write the number 5.
Step 2
Make a knight move in clockwise and write a number in the empty square such that the value of b3 + the value of the next move = 10.
Step 3
Continue the knight moves such that the current value plus the value of the next move is 10 until all knight moves are completed.

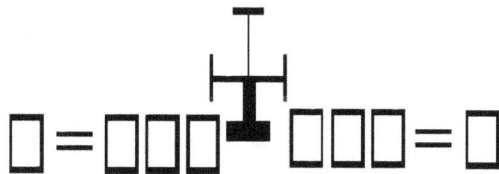

□ = □□□ ♘ □□□ = □

3

b

Knight moves to make 10

3	6	4	**6**
2	4		4
1	6	4	6
	a	**b**	**c**

Step 1
Start at c3 and write the number 6.
Step 2
Make a knight move in clockwise and write a number in the empty square such that the value of c3 + the value of the next move = 10.
Step 3
Continue the knight moves such that the current value plus the value of the next move is 10 until all knight moves are completed.

3

□ = □ □ □
□
□
||
□

c

Doubling, adding 10, or 9 + d

Students need to write the entire question before writing the result 10 to reinforce the idea what 2 numbers added together to make 10.

5		⦁			⦁	
4	⦁	⦁	⦁		⦁	⦁
3		⦁	2 3 1 / 2 3 2 / 3 2 3		⦁	
2	⦁	⦁	⦁		⦁	⦁
1		⦁			⦁	
	a	b	c		d	e

You are a chess piece located at c3 ▦.

⦁ = 1

Student Name _____　Date _____

Making 10 or adding around 10 or doubling

You are a chess piece located at c3.

● = 1

	2	4	3
	4	4	4
	3	4	2

Ho Math Chess 何数棋谜　妈！我会棋谜式加法啦！

Mom! I Learn Addition Using Math-Chess-Puzzles Connection

Contents include both traditional and Math-Chess-Puzzles combined methods. Extra strength

©2008 – 2018 Frank Ho, Amanda Ho　　All rights reserved. www.homathchess.com

Student Name _____ Date _____

Making 10 or adding around 10 or doubling

You are a chess piece located at c3 .

= 1

Student Name _____ Date _____

Making 10 or adding around 10 or doubling

You are a chess piece located at c3 ⊞ .

● = 1

Student Name _____ Date _____

Making 10 or adding around 10 or doubling

5					
4					
3			6 5 5 / 6 7 7 / 5 5 7		
2					
1					
	a	b	c	d	e

You are a chess piece located at c3 .

● = 1

Student Name _____ Date _____

Making 10 or adding around 10 or doubling

5		⠿		⠿	
4	⠿	⠿	⠿	⠿	⠿
3		⠿	7 7 7 / 8 8 8 / 6 8 6	⠿	
2	⠿	⠿	⠿	⠿	⠿
1		⠿		⠿	
	a	b	c	d	e

You are a chess piece located at c3 ▦.

⬤ = 1

Making 10 or adding around 10 or doubling

You are a chess piece located at c3 .

● = 1

**Only transfer each number to a correct position.
Do not the transfer the 3 by 3 square.**

Student Name _____ Date _____

Making 10 or adding around 10 or doubling

You are a chess piece located at c3 [grid].
● = 1

Only transfer each number to a correct position. Do not the transfer the 3 by 3 square.

Making 10 or adding around 10 or doubling

You are a chess piece located at c3.

● = 1

Only transfer each number to a correct position.
Do not the transfer the 3 by 3 square.

Making 10 or adding around 10 or doubling

You are a chess piece located at c3.

● = 1

Only transfer each number to a correct position.
Do not the transfer the 3 by 3 square.

Ho Math Chess 何数棋谜   妈！我会棋谜式加法啦！
Mom! I Learn Addition Using Math-Chess-Puzzles Connection
Contents include both traditional and Math-Chess-Puzzles combined methods. Extra strength
©2008 – 2018 Frank Ho, Amanda Ho   All rights reserved. www.homathchess.com
Student Name _____
                                                    Date _____

Making 10 or adding around 10 or doubling

You are a chess piece located at c3 .
● = 1

**Only transfer each number to a correct position.
Do not the transfer the 3 by 3 square.**

Making 10 or adding around 10 or doubling

You are a chess piece located at c3 .

● = 1

**Only transfer each number to a correct position.
Do not the transfer the 3 by 3 square.**

Student Name _____ Date _____

Making 10 or adding around 10 or doubling

You are a chess piece located at c3.

● = 1

**Only transfer each number to a correct position.
Do not the transfer the 3 by 3 square.**

Ho Math Chess 何数棋谜　妈！我会棋谜式加法啦！
Mom! I Learn Addition Using Math-Chess-Puzzles Connection
Contents include both traditional and Math-Chess-Puzzles combined methods. Extra strength

©2008 − 2018 Frank Ho, Amanda Ho　All rights reserved. www.homathchess.com

Student Name _____ Date _____

Making 10 or adding around 10 or doubling

5		●●●		●●●	
4	●●●	●●●	●●●	●●●	●●●
3		●●●	4 2 3 / 4 3 4 / 3 2 3	●●●	
2	●●●	●●●	●●●	●●●	●●●
1		●●●		●●●	
	a	b	c	d	e

You are a chess piece located at c3 .

● = 1

Only transfer each number to a correct position.
Do not the transfer the 3 by 3 square.

Mom! I Learn Addition Using Math-Chess-Puzzles Connection

Contents include both traditional and Math-Chess-Puzzles combined methods. Extra strength

Student Name _____ Date _____

Making 10 or adding around 10 or doubling

You are a chess piece located at c3 .

⬤ = 1

Only transfer each number to a correct position.
Do not the transfer the 3 by 3 square.

Making 10 or adding around 10 or doubling

You are a chess piece located at c3.

● = 1

Only transfer each number to a correct position.
Do not the transfer the 3 by 3 square.

Making 10 or adding around 10 or doubling

You are a chess piece located at c3.

● = 1

Only transfer each number to a correct position.
Do not the transfer the 3 by 3 square.

Making 10 or adding around 10 or doubling

You are a chess piece located at c3.

● = 1

Only transfer each number to a correct position.
Do not the transfer the 3 by 3 square.

Student Name _____　Date _____

Chess and math integrated problems

How many points altogether is the square ✖ or ☒ being attacked?

Answer _____

Answer _____

Answer _____

Answer _____

Ho Math Chess 何数棋谜　妈！我会棋谜式加法啦！
Mom! I Learn Addition Using Math-Chess-Puzzles Connection
Contents include both traditional and Math-Chess-Puzzles combined methods. Extra strength

©2008 – 2018 Frank Ho, Amanda Ho　All rights reserved. www.homathchess.com

Student Name _____ Date _____

How many points altogether is the square ✖ or ☒ being attacked?

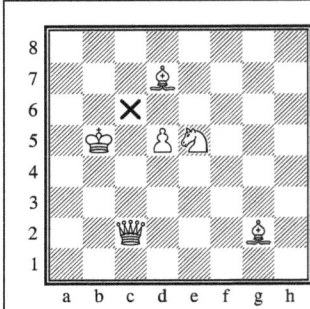

Answer _____

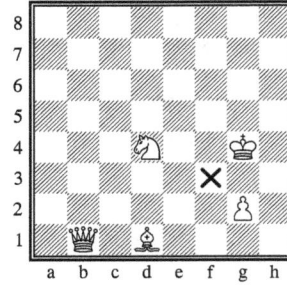

Answer _____

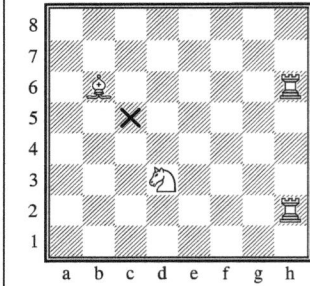

Answer _____

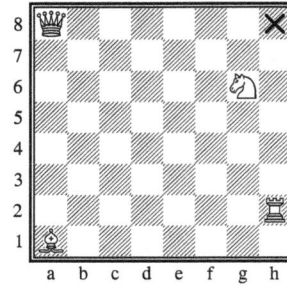

Answer _____

Student Name _____　Date _____

How many points altogether is the square ✖ or ☒ being attacked?

Answer _____

Answer _____

Answer _____

Answer _____

How many points altogether is the square 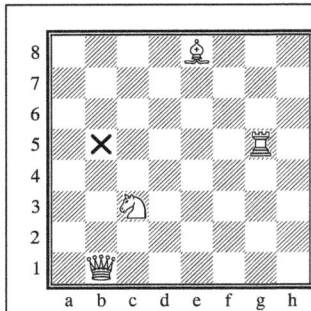 or 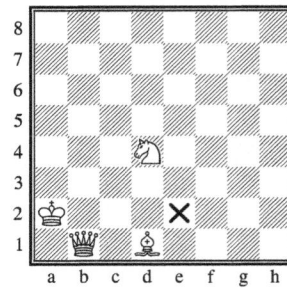 being attacked?

Answer _____

Answer _____

Answer _____

Answer _____

Student Name ______________________________ Date ____________________

8 plus

8	♙	2	8	♗
+ 1	+ 8	+ 8	+ 2	+ 8
8	4	8	♖	♖
+ 3	+ 8	+ 4	+ ♖	+ 8
8	6	6	8	7
+ 5	+ 6	+ 8	+ 6	+ 7

Ho Math Chess 何数棋谜　　妈！我会棋谜式加法啦！

Mom! I Learn Addition Using Math-Chess-Puzzles Connection

Contents include both traditional and Math-Chess-Puzzles combined methods. Extra strength

Student Name ______________________________ Date ____________________

7	7	8	♕	8
+ 8	+ 8	+ 8	+ 8	+ 9
6	6	8	5	8
+ 6	+ 8	+ 6	+ ♖	+ 5

7 plus

7	♙	2	7	7
+ 1	+ 7	+ 7	+ 2	+ 3
3	4	7	♖	5
+ 7	+ 7	+ 4	+ ♖	+ 7
7	6	6	7	7
+ 5	+ 6	+ 7	+ 6	+ 7
7	8	7	10	7
+ 8	+ 7	+ 7	+ 7	+ 9
♕	8	7	7	8
+ 7	+ 8	+ 8	+ 7	+ 7

Student Name _____ Date _____

6 plus

♙	6	2	6	♗
+ 6	+ ♙	+ 6	+ 2	+ 6
6	4	6	♖	♖
+ 3	+ 6	+ 4	+ 5	+ 6
6	6	6	7	6
+ 5	+ 6	+ 6	+ 6	+ 7
6	6	8	10	♕
+ 6	+ 8	+ 6	+ 6	+ 6
10	♕	6	6	8
+ 6	+ 6	+ 6	+ 8	+ 6

5 plus

1	♖	2	♖	3
+ 5	+ 1	+ 5	+ 2	+ 5
5	4	♖	4	♖
+ ♗	+ 4	+ 4	+ 5	+ ♖
♖	♖	6	7	7
+ 5	+ 6	+ 5	+ 5	+ 5
5	8	10	♕	♖
+ 8	+ 5	+ 5	+ 5	+ ♕
8	5	♖	7	♕
+ 5	+ 8	+ 7	+ 5	+ 5

Student Name _____ Date _____

4 plus

4 + ♟	♟ + 4	4 + 2	2 + 4	♝ + 4
4 + 3	4 + 4	4 + 5	♖ + 4	4 + 6
6 + 4	♝ + 7	7 + 3	7 + 4	4 + 7
♖ + 5	4 + 8	8 + 4	10 + 4	4 + ♛
♛ + 4	8 + 4	4 + 8	♛ + 4	4 + 9

Additions to 11

Fill in the blank circles with numbers.

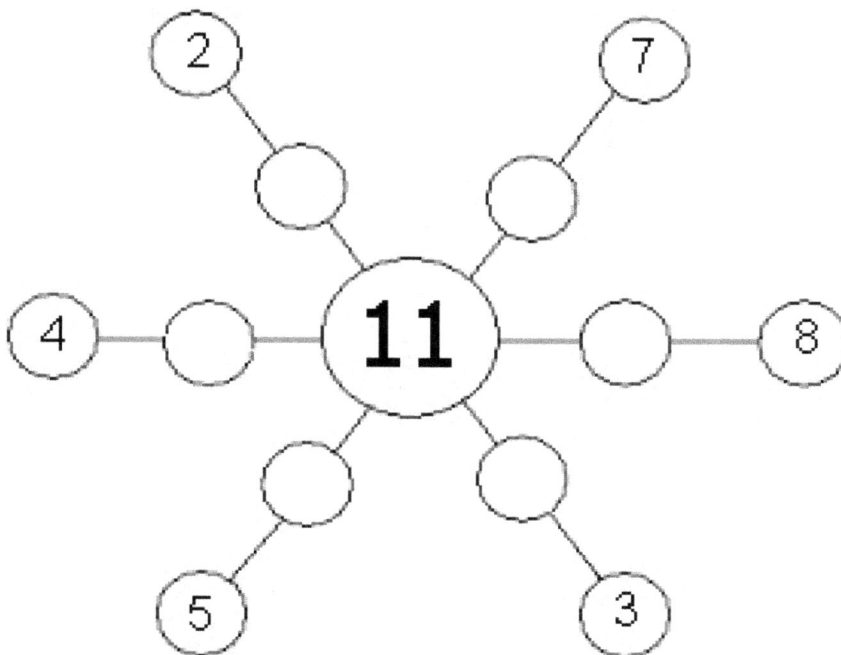

Additions to 12

Fill in the blank circles with numbers.

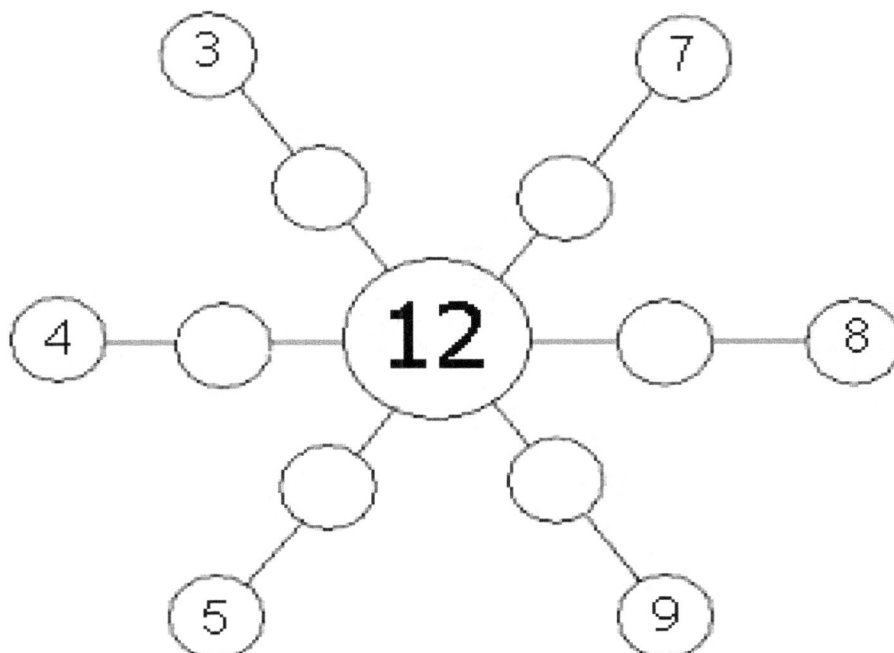

Additions to 13

Fill in the blank circles with numbers.

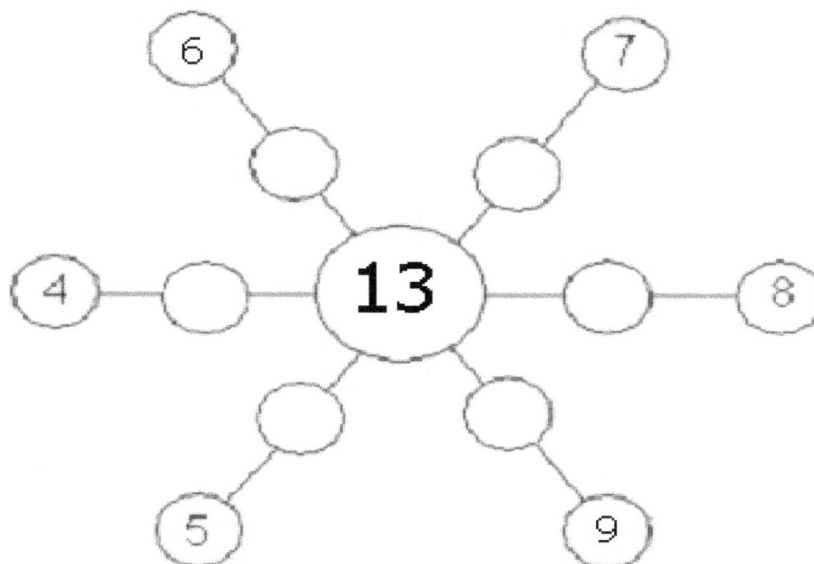

Additions to 14

Fill in the blank circles with numbers.

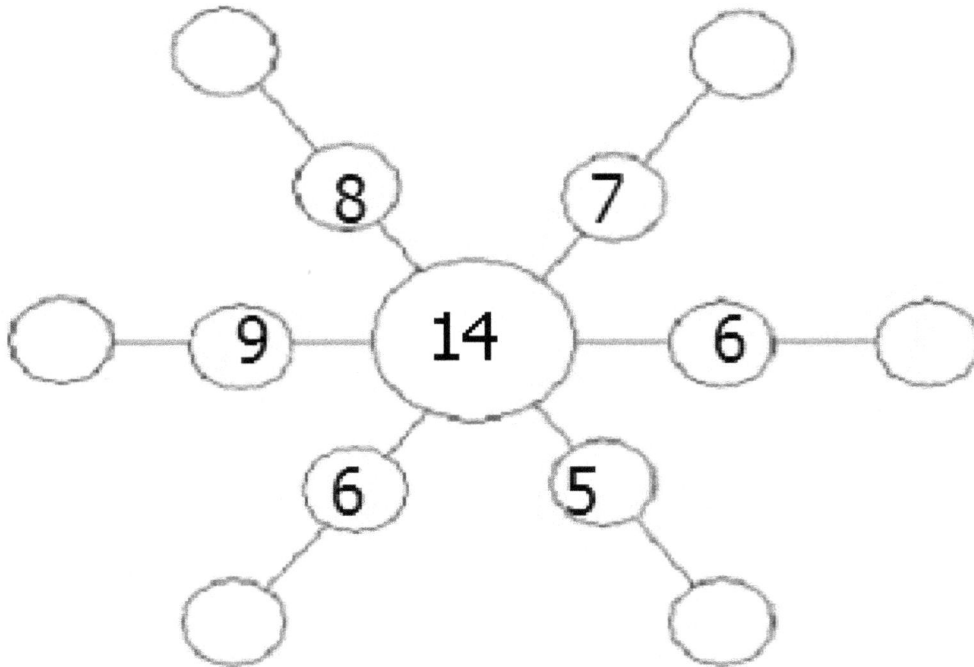

Additions to 15

Fill in the blank circles with numbers.

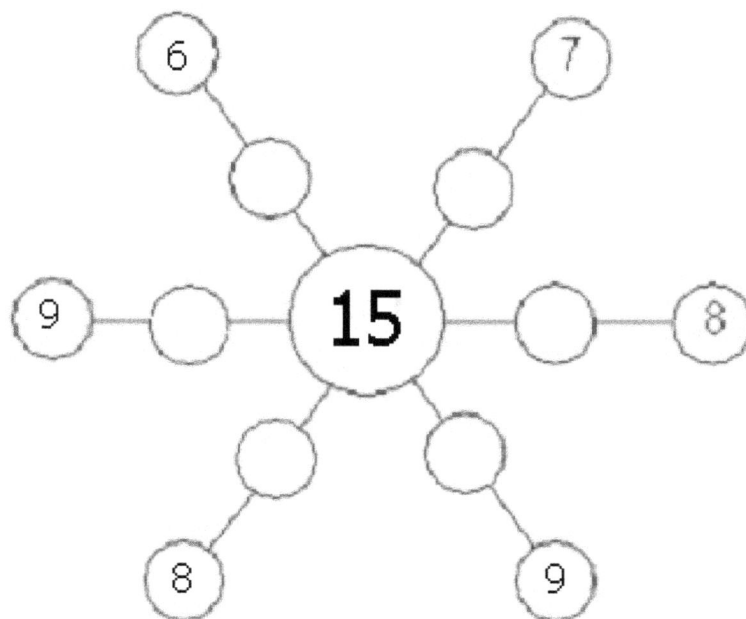

Additions to 16

Fill in the blank circles with numbers.

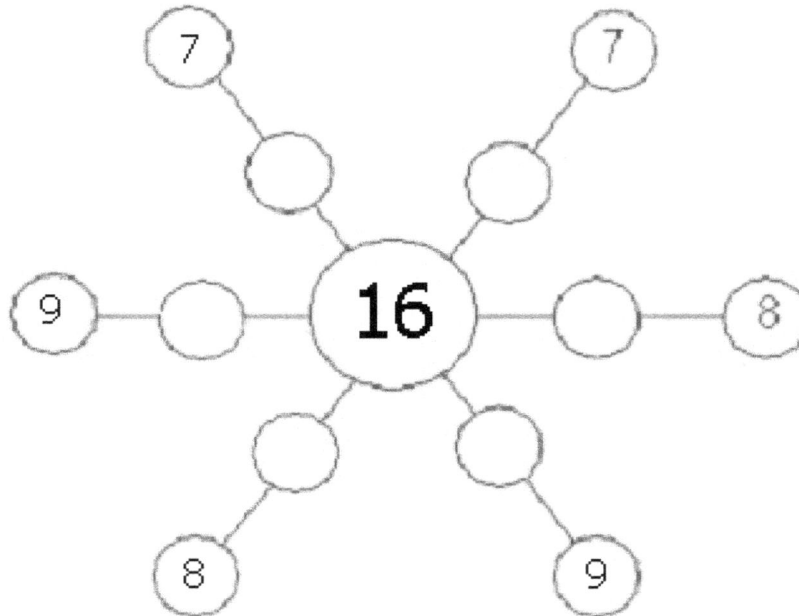

Using 8 to add

Fill in the blank circles with numbers.

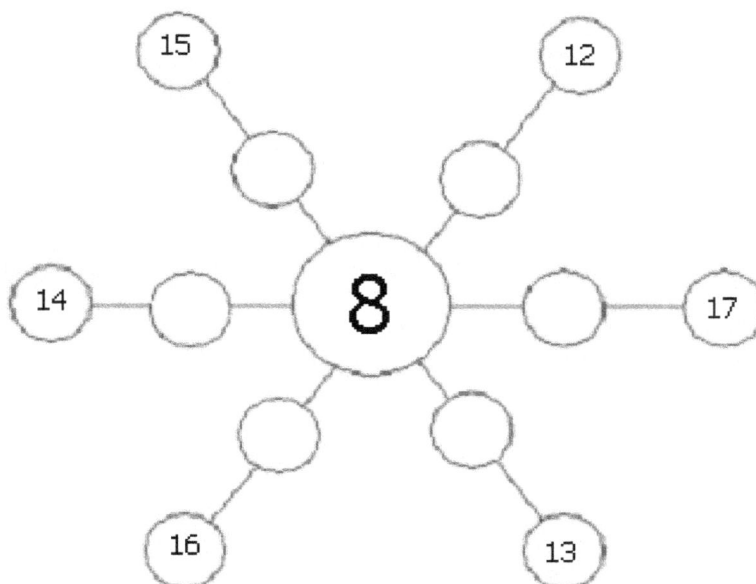

Student Name _____ Date _____

Using 9 to add

Fill in the blank circles with numbers.

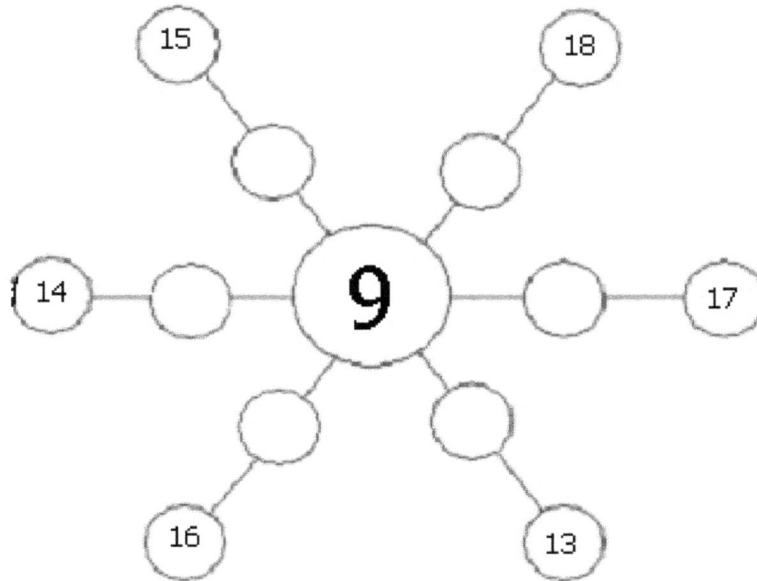

Ho Math Chess 何数棋谜　妈！我会棋谜式加法啦！
Mom! I Learn Addition Using Math-Chess-Puzzles Connection
Contents include both traditional and Math-Chess-Puzzles combined methods. Extra strength

©2008 – 2018 Frank Ho, Amanda Ho　　All rights reserved. www.homathchess.com

Student Name _____ Date _____

Number relationships

3	g1	g2	g3
2	h1		h2
1	i1	i2	i3
	a	b	c

3			
2			
1			
	d	e	f

3			
2			
1			
	g	h	i

You are at b2.

+ d1= _____ + _____ = _____

+ d2= _____ + _____ = _____

+ d3= _____ + _____ = _____

+ e1= _____ + _____ = _____

+ e2= _____ + _____ = _____

+ e3= _____ + _____ = _____

+ f1= _____ + _____ = _____

+ f2= _____ + _____ = _____

+ f3= _____ + _____ = _____

+e3+f2= ___ + ___ + ___ = _____

+d3+g2= ___ + ___ + ___ = _____

Student Name _____ Date _____

Number relationships

3	g2	g1	g3
2	h2		h1
1	i2	i1	i3
	a	b	c

3			
2			
1			
	d	e	f

3			
2			
1			
	g	h	i

You are at b2.

+ d1= _____ + _____ = _____

+ d2= _____ + _____ = _____

+ d3= _____ + _____ = _____

+ e1= _____ + _____ = _____

+ e2= _____ + _____ = _____

+ e3= _____ + _____ = _____

+ f1= _____ + _____ = _____

+ f2= _____ + _____ = _____

+ f3= _____ + _____ = _____

+e3+f2= ___ + ___ + ___ = _____

+d3+g2= ___ + ___ + ___ = _____

Student Name _____ Date _____

Number relationships

3	g3	g2	g1
2	h1		h2
1	i3	i2	i1
	a	b	c

3			
2			
1			
	d	e	f

3			
2			
1			
	g	h	i

You are at b2.

+ d1= _____ + _____ = _____

+ d2= _____ + _____ = _____

+ d3= _____ + _____ = _____

+ e1= _____ + _____ = _____

+ e2= _____ + _____ = _____

+ e3= _____ + _____ = _____

+ f1= _____ + _____ = _____

+ f2= _____ + _____ = _____

+ f3= _____ + _____ = _____

+e3+f2= ___ + ___ + ___ = _____

+d3+g2= ___ + ___ + ___ = _____

295

Student Name _____ Date _____

Number relationships

3	i1	i2	i3
2	g1		g2
1	h1	h2	h3
	a	b	c

3			
2			
1			
	d	e	f

3			
2			
1			
	g	h	i

You are at b2.

$+ d1=$ _____ + _____ = _____

$+ d2=$ _____ + _____ = _____

$+ d3=$ _____ + _____ = _____

$+ e1=$ _____ + _____ = _____

$+ e2=$ _____ + _____ = _____

$+ e3=$ _____ + _____ = _____

$+ f1=$ _____ + _____ = _____

$+ f2=$ _____ + _____ = _____

$+ f3=$ _____ + _____ = _____

$+e3+f2=$ ___ + ___ + ___ = _____

$+d3+g2=$ _____ + _____ + _____ = _____

Student Name _____ Date _____

Number relationships

3	i2	i1	i3
2	g2		g1
1	h2	h1	h3
	a	b	c

3			
2			
1			
	d	e	f

3			
2			
1			
	g	h	i

You are at b2.

+ d1= _____ + _____ = _____

+ d2= _____ + _____ = _____

+ d3= _____ + _____ = _____

+ e1= _____ + _____ = _____

+ e2= _____ + _____ = _____

+ e3= _____ + _____ = _____

+ f1= _____ + _____ = _____

+ f2= _____ + _____ = _____

+ f3= _____ + _____ = _____

+e3+f2= ___ + ___ + ___ = _____

+d3+g2= ___ + ___ + ___ = _____

Number relationships

3	i3	i2	i1
2	g1		g2
1	h3	h2	h1
	a	b	c

3			
2			
1			
	d	e	f

3			
2			
1			
	g	h	i

You are at b2.

+ d1= _____ + _____ = _____

+ d2= _____ + _____ = _____

+ d3= _____ + _____ = _____

+ e1= _____ + _____ = _____

+ e2= _____ + _____ = _____

+ e3= _____ + _____ = _____

+ f1= _____ + _____ = _____

+ f2= _____ + _____ = _____

+ f3= _____ + _____ = _____

+e3+f2= ___ + ___ + ___ = _____

+d3+g2= ____ + ____ + ____ = ____

Student Name _____ Date _____

Number relationships

3	h1	h2	h3
2	i1		i2
1	g1	g2	g3
	a	b	c

3			
2			
1			
	d	e	f

3			
2			
1			
	g	h	i

You are at b2.

+ d1 = _____ + _____ = _____

+ d2 = _____ + _____ = _____

+ d3 = _____ + _____ = _____

+ e1 = _____ + _____ = _____

+ e2 = _____ + _____ = _____

+ e3 = _____ + _____ = _____

+ f1 = _____ + _____ = _____

+ f2 = _____ + _____ = _____

+ f3 = _____ + _____ = _____

+ e3 + f2 = ___ + ___ + ___ = _____

+ d3 + g2 = ___ + ___ + ___ = _____

Student Name _____ Date _____

Number relationships

3	h2	h1	h3
2	i2		i1
1	g2	g1	g3
	a	b	c

3			
2			
1			
	d	e	f

3			
2			
1			
	g	h	i

You are at b2.

+ d1= _____ + _____ = _____

+ d2= _____ + _____ = _____

+ d3= _____ + _____ = _____

+ e1= _____ + _____ = _____

+ e2= _____ + _____ = _____

+ e3= _____ + _____ = _____

+ f1= _____ + _____ = _____

+ f2= _____ + _____ = _____

+ f3= _____ + _____ = _____

+e3+f2= ___ + ___ + ___ = _____

+d3+g2= _____ + _____ + _____ = _____

Student Name _____ Date _____

Number relationships

3	h3	h2	h1
2	i1		i2
1	g3	g2	g1
	a	b	c

3			
2			
1			
	d	e	f

3			
2			
1			
	g	h	i

You are at b2.

+ d1= _____ + _____ = _____

+ d2= _____ + _____ = _____

+ d3= _____ + _____ = _____

+ e1= _____ + _____ = _____

+ e2= _____ + _____ = _____

+ e3= _____ + _____ = _____

+ f1= _____ + _____ = _____

+ f2= _____ + _____ = _____

+ f3= _____ + _____ = _____

+e3+f2= ___ + ___ + ___ = _____

+d3+g2= ___ + ___ + ___ = _____

Student Name _____ Date _____

Number relationships

3	d1	d2	d3
2	e1		e2
1	f1	f2	f3
	a	b	c

3			
2			
1			
	d	e	f

3			
2			
1			
	g	h	i

You are at b2.

+ d1= _____   + _____ = _____

+ d2= _____   + _____ = _____

+ d3= _____   + _____ = _____

+ e1= _____   + _____ = _____

+ e2= _____   + _____ = _____

+ e3= _____   + _____ = _____

+ f1= _____   + _____ = _____

+ f2= _____   + _____ = _____

+ f3= _____   + _____ = _____

+e3+f2= ___ + ___ + ___ = _____

+d3+g2= ___ + ___ + ___ = _____

Number relationships

3	d2	d1	d3
2	e2		e1
1	f2	f1	f3
	a	b	c

3			
2			
1			
	d	e	f

3			
2			
1			
	g	h	i

You are at b2.

+ d1= _____ + _____ = _____

+ d2= _____ + _____ = _____

+ d3= _____ + _____ = _____

+ e1= _____ + _____ = _____

+ e2= _____ + _____ = _____

+ e3= _____ + _____ = _____

+ f1= _____ + _____ = _____

+ f2= _____ + _____ = _____

+ f3= _____ + _____ = _____

+e3+f2= ___ + ___ + ___ = _____

+d3+g2= ____ + ____ + ____ = _____

Ho Math Chess 何数棋谜 妈！我会棋谜式加法啦！

Mom! I Learn Addition Using Math-Chess-Puzzles Connection

Contents include both traditional and Math-Chess-Puzzles combined methods. Extra strength

Student Name _____ Date _____

Number relationships

3	d3	d2	d1
2	e1		e2
1	f3	f2	f1
	a	b	c

You are at b2.

$+ d1=$ _____ $+$ _____ $=$ _____

$+ d2=$ _____ $+$ _____ $=$ _____

$+ d3=$ _____ $+$ _____ $=$ _____

$+ e1=$ _____ $+$ _____ $=$ _____

$+ e2=$ _____ $+$ _____ $=$ _____

$+ e3=$ _____ $+$ _____ $=$ _____

$+ f1=$ _____ $+$ _____ $=$ _____

$+ f2=$ _____ $+$ _____ $=$ _____

$+ f3=$ _____ $+$ _____ $=$ _____

$+e3+f2=$ ___ $+$ ___ $+$ ___ $=$ _____

$+d3+g2=$ _____ $+$ ____ $+$ ____ $=$

Student Name _____ Date _____

Number relationships

3	f1	f2	f3
2	d1		d2
1	e1	e2	e3
	a	b	c

3			
2			
1			
	d	e	f

3			
2			
1			
	g	h	i

You are at b2.

+ d1= _____ + _____ = _____

+ d2= _____ + _____ = _____

+ d3= _____ + _____ = _____

+ e1= _____ + _____ = _____

+ e2= _____ + _____ = _____

+ e3= _____ + _____ = _____

+ f1= _____ + _____ = _____

+ f2= _____ + _____ = _____

+ f3= _____ + _____ = _____

+e3+f2= ___ + ___ + ___ = _____

+d3+g2= ___ + ___ + ___ = _____

Student Name _____ Date _____

Number relationships

3	f2	f1	f3
2	d2		d1
1	e2	e1	e3
	a	b	c

3			
2			
1			
	d	e	f

3			
2			
1			
	g	h	i

You are at b2.

+ d1= _____ + _____ = _____

+ d2= _____ + _____ = _____

+ d3= _____ + _____ = _____

+ e1= _____ + _____ = _____

+ e2= _____ + _____ = _____

+ e3= _____ + _____ = _____

+ f1= _____ + _____ = _____

+ f2= _____ + _____ = _____

+ f3= _____ + _____ = _____

+e3+f2= ___ + ___ + ___ = _____

+d3+g2= ___ + ___ + ___ = _____

Student Name _____ Date _____

Number relationships

3	f3	f2	f1
2	d2		d1
1	e2	e2	e1
	a	b	c

3			
2			
1			
	d	e	f

3			
2			
1			
	g	h	i

You are at b2.

+ d1= _____　+ _____ = _____

+ d2= _____　+ _____ = _____

+ d3= _____　+ _____ = _____

+ e1= _____　+ _____ = _____

+ e2= _____　+ _____ = _____

+ e3= _____　+ _____ = _____

+ f1= _____　+ _____ = _____

+ f2= _____　+ _____ = _____

+ f3= _____　+ _____ = _____

+e3+f2= ___ + ___ + ___ = _____

+d3+g2= ___ + ___ + ___ = _____

Number relationships

3	e1	e2	e3
2	d1		d2
1	f1	f2	f3
	a	b	c

3			
2			
1			
	d	e	f

3			
2			
1			
	g	h	i

You are at b2.

+ d1= _____ + _____ = _____

+ d2= _____ + _____ = _____

+ d3= _____ + _____ = _____

+ e1= _____ + _____ = _____

+ e2= _____ + _____ = _____

+ e3= _____ + _____ = _____

+ f1= _____ + _____ = _____

+ f2= _____ + _____ = _____

+ f3= _____ + _____ = _____

+e3+f2= ___ + ___ + ___ = _____

+d3+g2= ___ + ___ + ___ = _____

Number relationships

3	e2	e1	e3
2	d2		d1
1	f2	f1	f3
	a	b	c

3			
2			
1			
	d	e	f

3			
2			
1			
	g	h	i

You are at b2.

+ d1= _____   + _____ = _____

+ d2= _____   + _____ = _____

+ d3= _____   + _____ = _____

+ e1= _____   + _____ = _____

+ e2= _____   + _____ = _____

+ e3= _____   + _____ = _____

+ f1= _____   + _____ = _____

+ f2= _____   + _____ = _____

+ f3= _____   + _____ = _____

+e3+f2= ___ + ___ + ___ = _____

+d3+g2= ___ + ___ + ___ = _____

Student Name _____ Date _____

Number relationships

3	e3	e2	e1
2	d1		d2
1	f3	f2	f1
	a	b	c

You are at b2.

+ d1= _____   + _____ = _____

+ d2= _____   + _____ = _____

+ d3= _____   + _____ = _____

+ e1= _____   + _____ = _____

+ e2= _____   + _____ = _____

+ e3= _____   + _____ = _____

+ f1= _____   + _____ = _____

+ f2= _____   + _____ = _____

+ f3= _____   + _____ = _____

+e3+f2= ___ + ___ + ___ = _____

+d3+g2= _____ + _____ + _____ = _____

3			
2			
1			
	d	e	f

3			
2			
1			
	g	h	i

Student Name _____ Date _____

Number relationships

3	g2	g3	g1
2	h2		h1
1	I2	I3	I1
	a	b	c

3			
2			
1			
	d	e	f

3			
2			
1			
	g	h	i

You are at b2.

+ d1= _____ + _____ = _____

+ d2= _____ + _____ = _____

+ d3= _____ + _____ = _____

+ e1= _____ + _____ = _____

+ e2= _____ + _____ = _____

+ e3= _____ + _____ = _____

+ f1= _____ + _____ = _____

+ f2= _____ + _____ = _____

+ f3= _____ + _____ = _____

+e3+f2= ___ + ___ + ___ = _____

+d3+g2= ___ + ___ + ___ = _____

Student Name _____ Date _____

Number relationships

3	g1	g2	g3
2	h1		h2
1	i1	i2	i3
	a	b	c

3			
2			
1			
	d	e	f

3			
2			
1			
	g	h	i

You are at b2.

+ d1= _____ + _____ = _____

+ d2= _____ + _____ = _____

+ d3= _____ + _____ = _____

+ e1= _____ + _____ = _____

+ e2= _____ + _____ = _____

+ e3= _____ + _____ = _____

+ f1= _____ + _____ = _____

+ f2= _____ + _____ = _____

+ f3= _____ + _____ = _____

+e3+f2= ___ + ___ + ___ = _____

+d3+g2= ___ + ___ + ___ = _____

Student Name _____ Date _____

Number relationships

3	g1	g2	g3
2	h1		h2
1	i1	i2	i3
	a	b	c

	d	e	f
3			
2			
1			

	g	h	i
3			
2			
1			

You are at b2.

+ d1= _____   + _____ = _____

+ d2= _____   + _____ = _____

+ d3= _____   + _____ = _____

+ e1= _____   + _____ = _____

+ e2= _____   + _____ = _____

+ e3= _____   + _____ = _____

+ f1= _____   + _____ = _____

+ f2= _____   + _____ = _____

+ f3= _____   + _____ = _____

+e3+f2= ___ + ___ + ___ = _____

+d3+g2= ___ + ___ + ___ = _____

Student Name _____ Date _____

Number relationships

3	i1	i2	i3
2	d1		d2
1	g1	g2	g3
	a	b	c

3			
2			
1			
	d	e	f

3			
2			
1			
	g	h	i

You are at b2.

+ d1= _____ + _____ = _____

+ d2= _____ + _____ = _____

+ d3= _____ + _____ = _____

+ e1= _____ + _____ = _____

+ e2= _____ + _____ = _____

+ e3= _____ + _____ = _____

+ f1= _____ + _____ = _____

+ f2= _____ + _____ = _____

+ f3= _____ + _____ = _____

+e3+f2= ___ + ___ + ___ = _____

+d3+g2= ____ + ___ + ___ = _____

Student Name _____ Date _____

Number relationships

3	g1	g2	g3
2	h1		h2
1	i1	i2	i3
	a	b	c

3			
2			
1			
	d	e	f

3			
2			
1			
	g	h	i

You are at b2.

+ d3 = _____ + _____

+ d1 = _____ + _____

+ f2 = _____ + _____

+ e3 = _____ + _____

+ f2 = _____ + _____

+ d1 = _____ + _____

+ f2 = _____ + _____

+ f3 = _____ + _____

Part 4: Addition by carrying

d + dd

Wrong	Correct	Wrong	Correct
♖	₁5	7	₁7
+ 45	+ 45	+ 35	+ 35
4 1 0	5 0	3 1 2	4 2

6	7	♕	8
+ 66	+ 46	+ 16	+ 26
6	7	♕	♖
+ 55	+ 44	+ 33	+ 77
6	8	♕	4
+ 88	+ 48	+ 19	+ 26
6	♕	♖	8
+ 67	+ 46	+ 16	+ 26

Ho Math Chess 何数棋谜　妈！我会棋谜式加法啦！

Mom! I Learn Addition Using Math-Chess-Puzzles Connection

Contents include both traditional and Math-Chess-Puzzles combined methods. Extra strength

©2008 – 2018 Frank Ho, Amanda Ho　　All rights reserved. www.homathchess.com

Student Name _____　Date _____

Carrying or Regrouping for d + dd

6	4	7	♗	8
+1 4	+1 6	+1 3	+1 7	+1 2
2	♙	♕	♖	6
+1 8	+1 9	+1 1	+1 5	+1 4
4	7	8	6	8
+1 6	+1 3	+1 2	+1 4	+1 2
♕	♙	6	7	♗
+1 1	+1 9	+1 4	+1 3	+1 7
♖	6	8	♕	♙
+1 5	+1 4	+1 2	+1 1	+1 9

Carrying or Regrouping for d + dd

6	♖	7	4	8
+1 5	+1 6	+1 4	+1 7	+1 3
♗	2	♕	6	6
+1 8	+1 9	+1 2	+1 5	+1 5
6	7	8	♕	♖
+1 6	+1 7	+1 8	+1 9	+1 5
♕	4	♕	♕	♕
+1 3	+1 9	+1 5	+1 6	+1 7
♕	♕	8	7	6
+1 8	+1 9	+1 4	+1 4	+1 7

Carrying or Regrouping for d + dd

6 +3 4	4 +2 6	7 +2 3	♗ +3 7	8 +2 2
2 +2 8	♙ +3 9	♕ +2 1	♖ +3 5	6 +2 4
4 +3 6	7 +2 3	8 +3 2	6 +2 4	8 +3 2
♕ +3 1	♙ +2 9	6 +3 4	7 +2 3	■ +2 7
♖ +2 5	6 +3 4	8 +2 2	♕ +2 1	♙ +3 9

Carrying or Regrouping for d + dd

6	7	8	6	6
+1 6	+1 6	+1 6	+2 0	+1 9
7	7	♛	8	8
+1 7	+1 8	+1 7	+1 8	+1 9
8	♛	4	♜	♜
+1 8	+1 8	+1 4	+1 5	+1 4
♜	♜	8	♛	3
+1 6	+1 7	+1 5	+1 5	+1 7
7	7	8	8	4
+1 4	+1 4	+1 2	+1 3	+1 8

Ho Math Chess 何数棋谜　妈！我会棋谜式加法啦！
Mom! I Learn Addition Using Math-Chess-Puzzles Connection
Contents include both traditional and Math-Chess-Puzzles combined methods. Extra strength
©2008 – 2018 Frank Ho, Amanda Ho　All rights reserved. www.homathchess.com
Student Name _____
Date _____

Carrying or Regrouping for d + dd

6	4	7	3	8
+3 4	+5 6	+1 3	+4 7	+2 2
2	♙	♛	♜	6
+8 8	+7 9	+1 1	+2 5	+4 4
4	7	8	6	8
+6 6	+3 3	+5 2	+5 4	+8 2
♛	♙	6	7	■
+1 1	+4 9	+6 4	+6 3	+7 7
♜	6	8	♛	♙
+6 5	+3 4	+4 2	+5 1	+8 9

Student Name _____ Date _____

dd + d

1	1	1	1	1
17 + 6	18 + ♛	18 + 7	15 + 8	17 + ♜
27 + 8	28 + ♛	28 + 7	26 + ♜	25 + 7
39 + 8	35 + 6	36 + ♜	37 + 8	37 + ♜
49 + 8	47 + 6	48 + 7	49 + 8	48 + 6
59 + 8	55 + 6	57 + 6	56 + 7	56 + ♜

Student Name _____ Date _____

dd + d

5	7	1 7	2 7	3 7
+ 5	+ 5	+ 5	+ ♖	+ ♖
6	7	1 7	2 7	3 7
+ 6	+ 6	+ 6	+ 6	+ 6
7	7	1 7	2 7	3 7
+ 7	+ 7	+ 7	+ 7	+ 7
7	7	1 7	2 7	3 7
+ 7	+ 8	+ 8	+ 8	+ 8
10	7	1 7	2 7	3 7
+ 7	+ ♕	+ ♕	+ ♕	+ ♕

dd + d

♖	8	1 8	2 8	3 8
+ 5	+ ♖	+ ♖	+ 5	+ ♖
6	8	1 8	2 8	3 8
+ 6	+ 6	+ 6	+ 6	+ 6
7	8	1 8	2 8	3 8
+ 7	+ 7	+ 7	+ 7	+ 7
8	♕	1 8	2 8	3 8
+ 8	+ 8	+ 8	+ 8	+ 8
16	17	26	36	46 46
8	8	1 8	2 8	3 8
+ 8	+ ♕	+ ♕	+ ♕	+ 9

Ho Math Chess 何数棋谜　妈！我会棋谜式加法啦！
Mom! I Learn Addition Using Math-Chess-Puzzles Connection
Contents include both traditional and Math-Chess-Puzzles combined methods. Extra strength
©2008 – 2018 Frank Ho, Amanda Ho　　All rights reserved. www.homathchess.com
Student Name _____
Date _____

dd + d higher decades in pattern

5 + 5	♖ + 6	1 5 + 6	2 5 + 6	3 5 + 6
♖ + 6	1 5 + 6	2 5 + 6	3 5 + 6	4 5 + 6
1 5 + 6	2 5 + 6	3 5 + 6	4 5 + 6	5 5 + 6
2 5 + 6	3 5 + 6	4 5 + 6	5 5 + 6	6 5 + 6
3 5 + 6	4 5 + 6	5 5 + 6	6 5 + 6	7 5 + 6
4 5 + 6	5 5 + 6	6 5 + 6	7 5 + 6	+ 6

Student Name _____ Date _____

dd + d higher decades in pattern

5 + 5	5 + 7	1 5 + 7	2 5 + 7	3 5 + 7
♜ + 7	1 5 + 7	2 5 + 7	3 5 + 7	4 5 + 7
1 5 + 7	2 5 + 7	3 5 + 7	4 5 + 7	5 5 + 7
2 5 + 7	3 5 + 7	4 5 + 7	5 5 + 7	6 5 + 7
3 5 + 7	4 5 + 7	5 5 + 7	6 5 + 7	7 5 + 7
4 5 + 7	5 5 + 7	6 5 + 7	7 5 + 7	8 5 + 7

dd + d higher decades in pattern

5 + 5	♖ + 8	15 + 8	25 + 8	35 + 8
5 + 8	15 + 8	25 + 8	35 + 8	45 + 8
15 + 8	25 + 8	35 + 8	45 + 8	55 + 8
25 + 8	35 + 8	45 + 8	55 + 8	65 + 8
35 + 8	45 + 8	55 + 8	65 + 8	55 + 8
45 + 8	55 + 8	65 + 8	55 + 8	85 + 8

Student Name _____ Date _____

dd + d higher decades in pattern

5 + 5	5 + 9	1 5 + ♕	2 5 + 9	3 5 + ♕
5 + ♕	1 5 + 9	2 5 + 9	3 5 + ♕	4 5 + 9
1 5 + 9	2 5 + ♕	3 5 + 9	4 5 + ♕	5 5 + 9
2 5 + ♕	3 5 + 9	4 5 + ♕	5 5 + 9	9 5 + ♕
3 5 + 9	4 5 + ♕	5 5 + 9	6 5 + ♕	7 5 + 9
4 5 + 9	5 5 + ♕	6 5 + 9	7 5 + ♕	8 5 + 9

dd + d higher decades in pattern

6	7	1 7	2 7	3 7
+ 6	+ 6	+ 6	+ 6	+ 6
7	1 7	2 7	3 7	4 7
+ 6	+ 6	+ 6	+ 6	+ 6
1 7	2 7	3 7	4 7	5 7
+ 6	+ 6	+ 6	+ 6	+ 6
2 7	3 7	4 7	5 7	6 7
+ 6	+ 6	+ 6	+ 6	+ 6
3 7	4 7	5 7	6 7	7 7
+ 6	+ 6	+ 6	+ 6	+ 6
4 7	5 7	6 7	7 7	8 7
+ 6	+ 6	+ 6	+ 6	+ 6

dd + d higher decades in pattern

6 + 6	8 + 6	18 + 6	28 + 6	38 + 6
8 + 6	18 + 6	28 + 6	38 + 6	48 + 6
18 + 6	28 + 6	38 + 6	48 + 6	58 + 6
28 + 6	38 + 6	48 + 6	58 + 6	68 + 6
38 + 6	48 + 6	58 + 6	68 + 6	78 + 6
48 + 6	58 + 6	68 + 6	78 + 6	88 + 6

dd + d higher decades in pattern

6	♛	1 9	2 9	3 9
+ 6	+ 6	+ 6	+ 6	+ 6
9	1 9	2 9	3 9	4 9
+ 6	+ 6	+ 6	+ 6	+ 6
1 9	2 9	3 9	4 9	5 9
+ 6	+ 6	+ 6	+ 6	+ 6
2 9	3 9	4 9	5 9	6 9
+ 6	+ 6	+ 6	+ 6	+ 6
3 9	4 9	5 9	6 9	7 9
+ 6	+ 6	+ 6	+ 6	+ 6
4 9	5 9	6 9	7 9	8 9
+ 6	+ 6	+ 6	+ 6	+ 6

dd + d higher decades in pattern

7	7	1 7	2 7	3 7
+ 7	+ 8	+ 8	+ 8	+ 8
7	1 7	2 7	3 7	4 7
+ 8	+ 8	+ 8	+ 8	+ 8
1 7	2 7	3 7	4 7	5 7
+ 8	+ 8	+ 8	+ 8	+ 8
2 7	3 7	4 7	5 7	8 7
+ 8	+ 8	+ 8	+ 8	+ 8
3 7	4 7	5 7	6 7	7 7
+ 8	+ 8	+ 8	+ 8	+ 8
4 7	5 7	6 7	7 7	8 7
+ 8	+ 8	+ 8	+ 8	+ 8

Ho Math Chess 何数棋谜　妈！我会棋谜式加法啦！
Mom! I Learn Addition Using Math-Chess-Puzzles Connection
Contents include both traditional and Math-Chess-Puzzles combined methods. Extra strength
©2008 – 2018 Frank Ho, Amanda Ho　　All rights reserved. www.homathchess.com
Student Name _____
Date

dd + d higher decades in pattern

7 + 7	7 + ♛	17 + 9	27 + ♛	37 + 9
7 + ♛	17 + 9	27 + ♛	37 + 9	47 + ♛
17 + 9	27 + ♛	37 + 9	47 + ♛	57 + 9
27 + ♛	37 + 9	47 + ♛	57 + 9	97 + ♛
37 + 9	47 + ♛	57 + 9	67 + ♛	77 + 9
47 + 9	57 + ♛	67 + 9	77 + ♛	97 + 8

Ho Math Chess 何数棋谜　妈！我会棋谜式加法啦！

Mom! I Learn Addition Using Math-Chess-Puzzles Connection

Contents include both traditional and Math-Chess-Puzzles combined methods. Extra strength

©2008 – 2018 Frank Ho, Amanda Ho　　All rights reserved. www.homathchess.com

Student Name _____ Date _____

dd + d higher decades in pattern

8	8	1 8	2 8	3 8
+ 8	+ ♛	+ 9	+ ♛	+ 9
8	1 8	2 8	3 8	4 8
+ ♛	+ 9	+ ♛	+ 9	+ 9
1 8	2 8	3 8	4 8	5 8
+ 9	+ ♛	+ 9	+ ♛	+ 9
2 8	3 8	4 8	5 8	6 8
+ ♛	+ 9	+ ♛	+ 9	+ ♛
3 8	4 8	5 8	6 8	7 8
+ 9	+ ♛	+ 9	+ ♛	+ 9
4 8	5 8	6 8	7 8	8 8
+ ♛	+ 9	+ ♛	+ 9	+ ♛

Making multiples of 100

10 + 90	21 + 89	32 + 78	43 + 57	34 + 76
43 + 57	26 + 74	28 + 82	29 + 91	56 + 64
17 + 83	28 + 82	65 + 45	33 + 67	24 + 86
34 + 66	46 + 54	38 + 72	29 + 81	14 + 86
34 + 66	43 + 57	22 + 78	31 + 69	44 + 56
24 + 86	23 + 87	28 + 82	83 + 17	24 + 86

Student Name ______________________________ Date ____________________

dd + d

28	38	48	58	68
+ 9	+ ♕	+ 9	+ ♕	+ 9
38	48	58	68	78
+ ♕	+ 9	+ ♕	+ 9	+ ♕
48	58	68	78	88
+ 9	+ ♕	+ 9	+ ♕	+ 9

Ho Math Chess 何数棋谜　妈！我会棋谜式加法啦！

Mom! I Learn Addition Using Math-Chess-Puzzles Connection

Contents include both traditional and Math-Chess-Puzzles combined methods. Extra strength

©2008 − 2018 Frank Ho, Amanda Ho　　All rights reserved. www.homathchess.com

Student Name _____　Date _____

dd + d

67 + 6	69 + ♛	69 + 7	65 + 8	67 + ♜
77 + 8	78 + 9	78 + 7	76 + 5	75 + 7
89 + 8	85 + 6	86 + ♜	87 + 8	87 + 5
59 + 8	67 + 6	88 + 7	39 + 8	38 + 6
29 + 8	55 + 6	47 + 6	26 + 7	16 + ♜

dd + dd

11 + 11	22 + 22	33 + 33	44 + 44	10 + 10
20 + 20	30 + 30	40 + 40	10 + 20	20 + 30
30 + 20	40 + 50	12 + 28	17 + 23	14 + 26
15 + 25	26 + 34	13 + 27	18 + 22	19 + 21
35 + 45	46 + 36	37 + 47	48 + 38	39 + 49

Student Name _____ Date _____

Carrying or Regrouping for dd

Number profile	Standard Number	Write number in words
4 tens, 12 ones	⑤②	**Fifty-two**
4 tens, 10 ones	☐☐	
4 tens, 11 ones	☐☐	
9 tens, 19 ones	☐☐☐	
9 tens, 10 ones	☐☐☐	
7 tens, 13 ones	☐☐	
5 tens, 15 ones,	☐☐	
5 tens, 29 ones,	☐☐	
5 tens, 39 ones	☐☐	
1 ten, 69 ones	☐☐	
3 tens, 77 ones	☐☐☐	

dd + ddd adding 10

19 + 91	91 +119	28 + 82	28 +182	37 + 73
37 +173	46 + 64	46 +164	55 + 55	55 +155
28 + 82	28 +182	37 + 73	37 +173	19 + 91
91 +119	55 + 55	55 +155	46 + 64	46 +164
37 + 73	37 +173	28 + 82	28 +182	19 + 91

dd + ddd adding doubles

22 + 22	22 +122	33 + 33	33 +133	44 + 44
44 +144	66 +166	77 +177	55 + 55	55 +155
88 + 88	88 +188	99 + 99	99 +199	99 + 99
44 +144	55 + 55	55 +155	66 + 66	77 +177
77 + 77	77 +177	77 + 77	88 +188	88 + 88

dd + ddd adding 11

29 + 92	92 +129	38 + 83	38 +183	47 + 74
47 +174	56 + 65	56 +165	47 + 74	47 +174
38 + 83	38 +183	29 + 92	92 +129	56 + 65
56 +165	47 + 74	47 +174	56 + 65	56 +165
38 +183	38 + 83	29 + 92	92 +129	47 + 74

dd + ddd adding 12

39 + 93	93 +139	48 + 84	48 +184	57 + 75
66 +166	66 + 66	57 +175	48 + 84	39 +193
48 + 84	48 +184	39 + 93	93 +139	66 + 66
66 +166	57 + 75	57 +175	48 + 84	48 +184
48 +184	48 + 84	39 + 93	93 +139	57 + 75

dd + ddd adding 13

49 + 94	94 +149	58 + 85	58 +185	67 + 76
76 +167	67 + 76	67 +176	58 + 85	49 +194
58 + 85	58 +185	49 + 94	94 +149	76 + 67
67 +176	67 + 76	67 +176	58 + 85	58 +185
58 +185	58 + 85	49 + 94	94 +149	67 + 76

dd + ddd adding 14

59 + 95	95 +159	68 + 86	68 +186	77 + 77
86 +168	68 + 86	77 +177	68 + 86	59 +195
68 + 86	68 +186	59 + 95	95 +159	68 + 86
86 +168	77 + 77	77 +177	68 + 86	68 +186
68 +186	68 + 86	59 + 95	95 +159	77 + 77

dd + ddd adding 15

69 + 96	96 +169	78 + 87	78 +187	87 + 78
96 +169	69 + 96	87 +178	78 + 87	69 +196
78 + 87	78 +187	69 + 96	96 +169	69 + 96
96 +169	87 + 78	87 +178	78 + 87	78 +187
78 +187	78 + 87	69 + 96	96 +169	87 + 78

dd + ddd adding 16, 17, 18

79 + 97	97 +179	88 + 88	88 +188	97 + 79
89 +198	98 + 89	97 +179	99 + 99	99 +199
98 + 89	98 +189	99 + 99	99 +169	99 + 99
88 +188	97 + 79	97 +179	79 + 97	97 +179
88 +188	98 + 89	89 + 98	97 +179	97 + 79

dd + ddd

9 9 + 2 2	2 2 +1 9 9	9 9 + 3 3	3 3 +1 9 9	9 9 + 4 4
4 4 +1 9 9	9 9 + 1 1	1 1 +1 9 9	5 5 + 9 9	9 9 +1 5 5
6 6 + 9 9	7 7 +1 9 9	8 8 + 9 9	8 8 +1 9 9	6 6 + 9 9
6 6 +1 9 9	7 7 + 7 7	9 9 +1 9 9	9 9 + 8 8	4 4 +1 9 9
4 4 +1 9 9	9 9 + 8 8	7 7 + 9 9	7 7 +1 9 9	5 5 + 9 9

dd + ddd

88 + 33	33 +188	88 + 33	33 +188	88 + 44
44 +188	88 + 66	66 +188	55 + 88	88 +155
66 + 88	77 +188	88 + 88	88 +188	66 + 88
66 +188	77 + 77	99 +188	88 + 99	44 +188
44 +188	88 + 99	77 + 88	77 +188	55 + 88

dd + ddd

77 + 44	44 +177	77 + 44	55 +177	77 + 44
55 +177	77 + 66	66 +177	55 + 77	77 +155
66 + 77	77 +177	88 + 77	88 +177	66 + 77
66 +177	77 + 77	77 +199	99 + 77	44 +177
44 +177	77 + 88	77 + 77	77 +177	55 + 77

Ho Math Chess 何数棋谜　妈！我会棋谜式加法啦！

Mom! I Learn Addition Using Math-Chess-Puzzles Connection

Contents include both traditional and Math-Chess-Puzzles combined methods. Extra strength

©2008 – 2018 Frank Ho, Amanda Ho　All rights reserved. www.homathchess.com

Student Name _____ Date _____

dd + ddd

66 + 99	99 +166	66 + 88	88 +166	66 + 77
77 +166	66 + 55	55 +166	55 + 66	66 +155
66 + 66	77 +166	88 + 66	88 +166	66 + 66
66 +166	77 + 66	99 +166	66 + 88	77 +166
55 +166	66 + 88	77 + 66	77 +166	55 + 66

ddd + ddd

666	555	444	333	222
+766	+ 655	+544	+433	+322
777	888	999	666	555
+877	+ 988	+999	+766	+ 655
333	222	444	555	444
+433	+322	+544	+ 655	+544
666	555	777	888	999
+766	+ 655	+877	+ 988	+999
888	999	666	555	777
+ 988	+999	+766	+ 655	+877

Ho Math Chess 何数棋谜　妈！我会棋谜式加法啦！

Mom! I Learn Addition Using Math-Chess-Puzzles Connection

Contents include both traditional and Math-Chess-Puzzles combined methods. Extra strength

©2008 – 2018 Frank Ho, Amanda Ho　　All rights reserved. www.homathchess.com

Student Name _____

Date _____

ddd + ddd

766 +776	655 +665	544 +554	433 +443	322 +332
877 +887	988 +998	999 +999	766 +776	655 +665
766 +776	877 +887	544 +554	766 +776	988 +998
655 +665	544 +554	433 +443	322 +332	999 +999
766 +776	877 +887	544 +554	766 +776	988 +998

ddd + ddd

765 + 876	876 + 987	987 + 998	598 + 599	855 + 765
877 + 783	988 + 892	989 + 891	756 + 664	755 + 865
764 + 656	773 + 867	744 + 656	668 + 572	988 + 892
774 + 866	681 + 779	667 + 553	578 + 662	779 + 681
874 + 766	877 + 783	644 + 556	766 + 674	988 + 892

ddd + ddd

767	878	987	599	856
+ 876	+ 987	+ 999	+ 599	+ 765
877	988	989	756	755
+ 785	+ 895	+ 898	+ 668	+ 867
767	778	747	668	988
+ 656	+ 867	+ 656	+ 576	+ 897
779	684	667	578	779
+ 866	+ 779	+ 555	+ 665	+ 687
879	877	645	766	988
+ 766	+ 787	+ 556	+ 677	+ 897

Mom! I Learn Addition Using Math-Chess-Puzzles Connection

Contents include both traditional and Math-Chess-Puzzles combined methods. Extra strength

Student Name _____ Date _____

ddd + ddd

997 + 873	998 + 982	997 + 993	999 + 591	996 + 764
993 + 787	998 + 892	999 + 891	996 + 664	995 + 865
997 + 653	998 + 862	993 + 657	994 + 576	998 + 892
774 + 996	682 + 998	667 + 993	575 + 995	997 + 683
874 + 996	873 + 997	644 + 996	763 + 997	983 + 997

Student Name _____ Date _____

Mental addition

Example

100 + 200 + 1 = 1 *hundred* + 2 *hundred* and 1 more is 301.

Complete the following problems.

100 + 200 + 2 = ___ *hundred* + ___ *hundred* and ___ more is ____.
100 + 300 + 20 = ___ *hundred* + ___ *hundred* and ___ more is ____.
100 + 500 + 30 = ___ *hundred* + ___ *hundred* and ___ more is ____.
100 + 700 + 40 = ___ *hundred* + ___ *hundred* and ___ more is ____.
100 + 800 + 50 = ___ *hundred* + ___ *hundred* and ___ more is ____.
100 + 900 + 60 = ___ *hundred* + ___ *hundred* and ___ more is ____.
200 + 400 + 80 = ___ *hundred* + ___ *hundred* and ___ more is ____.
200 + 600 + 100 = ___ *hundred* + ___ *hundred* and ___ more is ____.
200 + 800 + 2 = ___ *hundred* + ___ *hundred* and ___ more is ____.

Mental addition

300 + 200 + 1 = 3 *hundred* + 2 *hundred* and 1 more is 501.

400 + 200 + 2 = ___ *hundred* + ___ *hundred* and ___ more is ____.

500 + 300 + 20 = ___ *hundred* + ___ *hundred* and ___ more is ____.

600 + 500 + 30 = ___ *hundred* + ___ *hundred* and ___ more is ____.

700 + 700 + 40 = ___ *hundred* + ___ *hundred* and ___ more is ____.

800 + 800 + 50 = ___ *hundred* + ___ *hundred* and ___ more is ____.

900 + 900 + 60 = ___ *hundred* + ___ *hundred* and ___ more is ____.

1000 + 500 + 80 = ___ thousand + ___ *hundred* and ___ more is ____.

1000 + 500 + 100 = ___ thousand + ___ *hundred* + ___ *hundred* is ____.

1000 + 600 +190 = ___ thousand + ___ *hundred* + ___ *hundred* and ___ more is ____.

1000 + 700 +190 = ___ thousand + ___ *hundred* + ___ *hundred* and ___ more is ____.

1000 + 800 +90 = ___ thousand + ___ *hundred* + ___ *hundred* and ___ more is ____.

d + d + d

7	1	2	♖	6
♗	♖	2	4	6
+ 4	+ 9	+ 8	+ 5	+ 4

7	5	9	8	7
3	1	2	4	6
+ ♗	+ 9	+ 8	+ 6	+ 4

7	1	8	6	6
3	9	2	4	♗
+ 7	+ 9	+ 8	+ 5	+ 7

7	9	9	7	6
4	1	2	5	5
+ 6	+ 9	+ 8	+ ♖	+ 4

d + d + d

6	8	2	♖	4
♗	♖	5	4	6
+ 4	+ 2	+ 8	+ 6	+ 4

7	1	9	8	7
4	2	1	2	6
+ ♗	+ 9	+ 8	+ 6	+ 3

7	2	8	5	6
3	9	4	5	♗
+ 8	+ 9	+ 8	+ 5	+ 7

6	9	9	5	6
4	9	1	5	6
+ 6	+ 9	+ 8	+ ♖	+ 4

d + d + d

7	2	9	♖	8
♗	♖	2	9	6
+ 8	+ 8	+ 8	+ 5	+ 4

7	1	9	3	9
7	1	1	4	6
+ ♗	+ 9	+ 8	+ 7	+ 4

7	1	8	6	7
3	1	2	4	♗
+ 3	+ 9	+ 2	+ 4	+ 7

7	8	9	7	6
3	2	1	5	4
+ 3	+ 2	+ 9	+ ♖	+ 4

d + d + d

6 ♗ + 4	♖ ♖ + 9	3 2 + 8	♖ 4 + ♖	6 ♖ + 4
♖ 3 + ♗	5 1 + ♕	♕ 2 + 8	♗ 4 + 6	♕ 6 + 4
7 4 + 7	2 9 + 9	8 2 + 3	6 5 + ♖	7 ♗ + 7
7 5 + 6	9 2 + 9	9 2 + 9	6 5 + ♖	6 5 + 5

Number pattern

1. ♙ , 4, 7, 10, 13, 16, □, □, □, □

2. ♗ , 6, ♕ , 12, 15, □, □, □, □

3. 2, ♖ , 8, □, □, □, □

4. ♗ , ♖ , 7, □, □, □, □

5. 99, 96, 93, 90, 87, □, □, □, □

6. 105, 100, 95, 90, □, □, □, □

7. 100, 90, 80, 70, □, □, □, □

8. 104, 100, 96, 92, □, □, □, □

Number pattern

1. ☐, ☐, ☐, ☐, 15, 13, 11, ♛, 7, ♜

2. ♝, 6, ♛, 12, 15, ☐, ☐, ☐, ☐

3. 15, 13, 11, ♛, ☐, ☐, ☐, ☐

4. 3, ♜, 7, ☐, ☐, ☐, ☐

5. ♚, ♜, 10, ☐, ☐, ☐, ☐

6. 98, 96, 94, 92, 90, ☐, ☐, ☐, ☐

7. 115, 110, 105, 100, ☐, ☐, ☐, ☐

8. 101, 91, 81, 71, ☐, ☐, ☐, ☐

9. 107, 109, 111, 113, ☐, ☐, ☐, ☐

Ho Math Chess 何数棋谜　妈！我会棋谜式加法啦！

Mom! I Learn Addition Using Math-Chess-Puzzles Connection

Contents include both traditional and Math-Chess-Puzzles combined methods. Extra strength

Student Name _____ Date _____

Number pattern

1. ♙+1, 7+1, 13+1, □, □, □, □

2. ♗-1, 6-1, ♛-1, □, □, □, □

3. 2+1, ♜, 7, ♛, □, □, □

4. ♗+1, ♜+1, 7+1, □, □, □, □

5. 99-2, 96-2, 93-2, 90-2, 87-2, □, □, □, □

6. 105-3, 100-3, 95-3, 90-3, □, □, □, □

7. 100-5, 90-5, 80-5, 70-5, □, □, □, □

8. 104+1, 100+1, 96+1, 92+1, □, □, □, □

High performance multi-digit addition

3	889	988	898
2	999	**333**	989
1	998	888	899
	a	b	c

The original square is at b2 = ☐.

$$090809$$
$$+\ 930303$$

b2 + ⬍➤ =
☐☐☐
+ ○○○

☐ + ✕ =
☐☐☐
+ ○○○

b2 + ⬍➤ =
☐☐☐
+ ○○○

☐ + ⬍➤ = 1332
☐☐☐
+ ○○○

b2 + ✕ =
☐☐☐
+ ○○○

☐ + ⬍➤ =
☐☐☐
+ ○○○

b2 + ✕ =
☐☐☐
+ ○○○

☐ + ✕ =
☐☐☐
+ ○○○

High performance multi-digit addition

3	778	877	787
2	888	**444**	878
1	887	888	788
	a	b	c

The original square is at b2 = ☐.

09080907
+ 94040404

b2 + ⬌ =
☐☐☐
+ ○○○

☐ + ✗ =
☐☐☐
+ ○○○

b2 + ⬌ =
☐☐☐
+ ○○○

☐ + ⬌ =
☐☐☐
+ ○○○

b2 + ✗ =
☐☐☐
+ ○○○

☐ + ⬌ =
☐☐☐
+ ○○○

b2 + ✗ =
☐☐☐
+ ○○○

☐ + ✗ =
☐☐☐
+ ○○○

Student Name _____ Date _____

High performance multi-digit addition

3	666	677	676
2	777	**555**	787
1	998	888	899
	a	b	c

The original square is at b2 = ☐ .

High performance multi-digit addition

3	889	988	898
2	999	**666**	989
1	998	888	899
	a	b	c

The original square is at b2 = ☐.

0908090607
+ 9606060606

b2 + ⬌⬍ =
☐☐☐
+ ○○○

☐ + ✕ =
☐☐☐
+ ○○○

b2 + ⬌⬍ =
☐☐☐
+ ○○○

☐ + ⬌⬍ =
☐☐☐
+ ○○○

b2 + ✕ =
☐☐☐
+ ○○○

☐ + ⬌⬍ =
☐☐☐
+ ○○○

b2 + ✕ =
☐☐☐
+ ○○○

☐ + ✕ =
☐☐☐
+ ○○○

High performance multi-digit addition

3	667	766	567
2	657	**777**	756
1	998	888	899
	a	b	c

The original square is at b2 = ☐.

090809050607
+ 970707070707

b2 + ⬌⬍(anchor) =
☐☐☐
+ ○○○
___

☐ + ✕ =
☐☐☐
+ ○○○
___

b2 + ⬌⬍(anchor) =
☐☐☐
+ ○○○
___

☐ + ⬌⬍(anchor) =
☐☐☐
+ ○○○
___

b2 + ✕ =
☐☐☐
+ ○○○
___

☐ + ⬌⬍(anchor) =
☐☐☐
+ ○○○
___

b2 + ✕ =
☐☐☐
+ ○○○
___

☐ + ✕ =
☐☐☐
+ ○○○
___

High performance multi-digit addition

3	344	454	565
2	563	**888**	678
1	789	987	976
	a	b	c

The original square is at b2 = ☐ .

0809070506040302
+ 9808080808080808

b2 + ⊕ =
☐☐☐
+ ○○○

☐ + ✕ =
☐☐☐
+ ○○○

b2 + ⊕ =
☐☐☐
+ ○○○

☐ + ⊕ =
☐☐☐
+ ○○○

b2 + ✕ =
☐☐☐
+ ○○○

☐ + ⊕ =
☐☐☐
+ ○○○

b2 + ✕ =
☐☐☐
+ ○○○

☐ + ✕ =
☐☐☐
+ ○○○

Computing through function concept

3	↓	2	⊥
2	4	↓	↔
1	6	7	8
	a	b	c

The original square is at b2.

a3 + b2 = ___ + ___ = ___
↓ + ↓ = 1 + 1 = 2

b3 + b2 = ___ + ___ = ___

c3 + b2 = ___ + ___ = ___

a2 + b2 = ___ + ___ = ___

b2 + b2 = ___ + ___ = ___

c2 + b2 = ___ + ___ = ___

a1 + b2 = ___ + ___ = ___

b1 + b2 = ___ + ___ = ___

c1 + b2 = ___ + ___ = ___

a3 − b2 = ___ − ___ = ___

b3 − b2 = ___ − ___ = ___

c3 − b2 = ___ − ___ = ___

a2 − b2 = ___ − ___ = ___

b2 − b2 = ___ − ___ = ___

c2 − b2 = ___ − ___ = ___

a1 − b2 = ___ − ___ = ___

b1 − b2 = ___ − ___ = ___

c1 − b2 = ___ − ___ = ___

Ho Math Chess 何数棋谜　妈！我会棋谜式加法啦！
Mom! I Learn Addition Using Math-Chess-Puzzles Connection
Contents include both traditional and Math-Chess-Puzzles combined methods. Extra strength
©2008 – 2018 Frank Ho, Amanda Ho　　All rights reserved. www.homathchess.com
Student Name _____
Date _____

From fingering math to computing through function concept

3	↓	2	⊢┬
2	4	2	↔
1	6	7	8
	a	b	c

The original square is at b2.

a3 + b2 = ___ + ___ = ___	b3 + b2 = ___ + ___ = ___
c3 + b2 = ___ + ___ = ___	a2 + b2 = ___ + ___ = ___
b2 + b2 = ___ + ___ = ___	c2 + b2 = ___ + ___ = ___
a1 + b2 = ___ + ___ = ___	b1 + b2 = ___ + ___ = ___
c1 + b2 = ___ + ___ = ___	b2 – a3 = ___ – ___ = ___
b3 – b2 = ___ – ___ = ___	c3 – b2 = ___ – ___ = ___
a2 – b2 = ___ – ___ = ___	b2 – b2 = ___ – ___ = ___
c2 – b2 = ___ – ___ = ___	a1 – b2 = ___ – ___ = ___
b1 – b2 = ___ – ___ = ___	c1 – b2 = ___ – ___ = ___

From fingering math to computing through function concept

3	↓	2	⊢⊤
2	5	⊢⊤	⟷
1	6	7	8
	a	b	c

The original square is at b2.

a3 + b2 = ___ + ___ = ___	b3 + b2 = ___ + ___ = ___
c3 + b2 = ___ + ___ = ___	a2 + b2 = ___ + ___ = ___
b2 + b2 = ___ + ___ = ___	c2 + b2 = ___ + ___ = ___
a1 + b2 = ___ + ___ = ___	b1 + b2 = ___ + ___ = ___
c1 + b2 = ___ + ___ = ___	b2 − a3 = ___ − ___ = ___
b2 − b3 = ___ − ___ = ___	c3 − b2 = ___ − ___ = ___
a2 − b2 = ___ − ___ = ___	b2 − b2 = ___ − ___ = ___
c2 − b2 = ___ − ___ = ___	a1 − b2 = ___ − ___ = ___
b1 − b2 = ___ − ___ = ___	c1 − b2 = ___ − ___ = ___

Ho Math Chess 何数棋谜　妈！我会棋谜式加法啦！
Mom! I Learn Addition Using Math-Chess-Puzzles Connection

Student Name _____ Date _____

From fingering math to computing through function concept

3	↓	2	⊢┬
2	4	4	⬅↕
1	6	7	8
	a	b	c

The original square is at b2.

a3 + b2 = ___ + ___ = ___	b3 + b2 = ___ + ___ = ___
c3 + b2 = ___ + ___ = ___	a2 + b2 = ___ + ___ = ___
b2 + b2 = ___ + ___ = ___	c2 + b2 = ___ + ___ = ___
a1 + b2 = ___ + ___ = ___	b1 + b2 = ___ + ___ = ___
c1 + b2 = ___ + ___ = ___	b2 − a3 = ___ − ___ = ___
b2 − b3 = ___ − ___ = ___	b2 − c3 = ___ − ___ = ___
a2 − b2 = ___ − ___ = ___	b2 − b2 = ___ − ___ = ___
c2 − b2 = ___ − ___ = ___	a1 − b2 = ___ − ___ = ___
b1 − b2 = ___ − ___ = ___	c1 − b2 = ___ − ___ = ___

From fingering math to computing through function concept

3	↔	5	↔
2	6	↔	↔
1	6	7	8
	a	b	c

The original square is at b2.

a3 + b2 = ___ + ___ = ___	b3 + b2 = ___ + ___ = ___
c3 + b2 = ___ + ___ = ___	a2 + b2 = ___ + ___ = ___
b2 + b2 = ___ + ___ = ___	c2 + b2 = ___ + ___ = ___
a1 + b2 = ___ + ___ = ___	b1 + b2 = ___ + ___ = ___
c1 + b2 = ___ + ___ = ___	a3 – b2 = ___ – ___ = ___
b3 – b2 = ___ – ___ = ___	c3 – b2 = ___ – ___ = ___
a2 – b2 = ___ – ___ = ___	b2 – b2 = ___ – ___ = ___
c2 – b2 = ___ – ___ = ___	a1 – b2 = ___ – ___ = ___
b1 – b2 = ___ – ___ = ___	c1 – b2 = ___ – ___ = ___

From fingering math to computing through function concept

3	7	7	8
2	8	6	✳
1	6	7	8
	a	b	c

The original square is at b2.

a3 + b2 = ___ + ___ = ___	b3 + b2 = ___ + ___ = ___
c3 + b2 = ___ + ___ = ___	a2 + b2 = ___ + ___ = ___
b2 + b2 = ___ + ___ = ___	c2 + b2 = ___ + ___ = ___
a1 + b2 = ___ + ___ = ___	b1 + b2 = ___ + ___ = ___
c1 + b2 = ___ + ___ = ___	a3 − b2 = ___ − ___ = ___
b3 − b2 = ___ − ___ = ___	c3 − b2 = ___ − ___ = ___
a2 − b2 = ___ − ___ = ___	b2 − b2 = ___ − ___ = ___
c2 − b2 = ___ − ___ = ___	a1 − b2 = ___ − ___ = ___
b1 − b2 = ___ − ___ = ___	c1 − b2 = ___ − ___ = ___

Student Name _____ Date _____

From fingering math to computing through function concept

3	7	✳	8
2	8	7	8
1	✳	7	8
	a	b	c

The original square is at b2.

a3 + b2 = ___ + ___ = ___	b3 + b2 = ___ + ___ = ___
c3 + b2 = ___ + ___ = ___	a2 + b2 = ___ + ___ = ___
b2 + b2 = ___ + ___ = ___	c2 + b2 = ___ + ___ = ___
a1 + b2 = ___ + ___ = ___	b1 + b2 = ___ + ___ = ___
c1 + b2 = ___ + ___ = ___	a3 − b2 = ___ − ___ = ___
b3 − b2 = ___ − ___ = ___	c3 − b2 = ___ − ___ = ___
a2 − b2 = ___ − ___ = ___	b2 − b2 = ___ − ___ = ___
c2 − b2 = ___ − ___ = ___	a1 − b2 = ___ − ___ = ___
b1 − b2 = ___ − ___ = ___	c1 − b2 = ___ − ___ = ___

Mom! I Learn Addition Using Math-Chess-Puzzles Connection

Contents include both traditional and Math-Chess-Puzzles combined methods. Extra strength

Student Name _____ Date _____

Computing additions through math and chess integrated puzzle

	a	b	c	d	e
5			g4		
4	g3	⊢	⊣	⊢	f3
3	g1	⊢	✕	⊢	f1
2	g2	⊢	⊣	⊢	f2
1			f4		

	f	g	h	i	j
5	✕	5	✸	9	✸
4	✸	✕	6	7	11
3	↓	2	⊢	✕	12
2	4	⊣		14	✕
1	✛	7	8	✕	15

The start square is at c3 = ☐.

c3 + ✛ = __ + __ = __	c3 + ✕ = __ + __ = __
☐ + ✛ = __ + __ = __	☐ + ✕ = __ + __ = __
c3 + ✛ = __ + __ = __	c3 + ✕ = __ + __ = __
☐ + ✛ = __ + __ = __	☐ + ✕ = __ + __ = __

Ho Math Chess 何数棋谜 妈！我会棋谜式加法啦！
Mom! I Learn Addition Using Math-Chess-Puzzles Connection
Contents include both traditional and Math-Chess-Puzzles combined methods. Extra strength

©2008 — 2018 Frank Ho, Amanda Ho   All rights reserved. www.homathchess.com

Student Name _____   Date _____

Computing additions through math and chess integrated puzzle

5		g4		
4	g3	⊥	⊤	⊥	f3
3	g1	⊥	✳	⊥	f1
2	g2	⊥	⊤	⊥	f2
1		f4		
	a	b	c	d	e

5	⤢	5	✳	9	✳
4	✳	⤢	6	7	11
3	⤓	2	⊥	⤢	12
2	4		⊥		14	⤢
1	⬌	7	8	⤢	15
	f	g	h	i	j

The start square is at c3 = ☐.

c3 + ⬌ = __ + __ = __	c3 + ⤢ = __ + __ = __
☐ + ⬌ = __ + __ = __	☐ + ⤢ = __ + __ = __
c3 + ⬌ = __ + __ = __	c3 + ⤢ = __ + __ = __
☐ + ⬌ = __ + __ = __	☐ + ⤢ = ___ + ___ = ___

Mom! I Learn Addition Using Math-Chess-Puzzles Connection

Contents include both traditional and Math-Chess-Puzzles combined methods. Extra strength

Student Name _____ Date _____

Computing additions through math and chess integrated puzzle

5		f2	f3	g2	f1
4		⊢⊤	⊢⊤	⊢⊤	
3		⊢⊤	↔	⊢⊤	
2		⊢⊤	⊢⊤	⊢⊤	
1	g1	f4	g3	g4	
	a	b	c	d	e

5	⤢	5	✳	9	✳
4	✳	⤢	6	7	11
3	↓	2	⊢⊤	⤢	12
2	4	⊢⊤		14	⤢
1	↔	7	8	⤢	15
	f	g	h	i	j

The start square is at c3 = ☐.

c3 + ↔ = ___ + ___ = ___

☐ + ↔ = ___ + ___ = ___

c3 + ⤓ = ___ + ___ = ___

☐ + ⤓ = ___ + ___ = ___

c3 + ⤢ = ___ + ___ = ___

☐ + ⤢ = ___ + ___ = ___

c3 + ⤢ = ___ + ___ = ___

☐ + ⤢ = ___ + ___ = ___

Computing additions through math and chess integrated puzzle

5		f2	f3	g2	f1
4		⊢	⊢	⊢	
3		⊢	⊠	⊢	
2		⊢	⊢	⊢	
1	g1	f4	g3	g4	
	a	b	c	d	e

5	⤢	5	✳	9	✳
4	✳	⤢	6	7	11
3	⤓		⊢	⤢	12
2	4	⊢		14	⤢
1	⬄	7	8	⤢	15
	f	g	h	i	j

The start square is at c3 = ☐.

c3 + ⬄ = __ + __ = __	c3 + ⤢ = __ + __ = __
☐ + ⬄ = __ + __ = __	☐ + ⤢ = __ + __ = __
c3 + ⬄ = __ + __ = __	c3 + ⤢ = __ + __ = __
☐ + ⬄ = __ + __ = __	☐ + ⤢ = __ + __ = __

Computing additions through math and chess integrated puzzle

5		f2	f3	g2	f1
4					
3					
2					
1	g1	f4	g3	g4	
	a	b	c	d	e

5		5		9	
4			6	7	11
3		2			12
2	4			14	
1		7	8		15
	f	g	h	i	j

The start square is at c3 = ☐.

c3 + ✛ = __ + __ = __

☐ + ✛ = __ + __ = __

c3 + ✛ = __ + __ = __

☐ + ✛ = __ + __ = __

c3 + ✛ = __ + __ = __

☐ + ✛ = __ + __ = __

c3 + ✕ = ___ + ___ = ___

☐ + ✕ = ___ + ___ = ___

c3 + ✕ = ___ + ___ = ___

☐ + ✕ = ___ + ___ = ___

c3 + ✕ = ___ + ___ = ___

☐ + ✕ = ___ + ___ = ___

Student Name _____ Date _____

Computing additions through math and chess integrated puzzle

5			g4	
4	g3	⊥	+	⊥	f3
3	g1	⊥	⬌	⊥	f1
2	g2	⊥	⊥	⊥	f2
1			f4		
	a	b	c	d	e

5	⤢	5	❋	9	❋
4	❋	⤢	6	7	11
3	⤓	2	+	⤢	12
2	4	+		14	⤢
1	⬌	7	8	⤢	15
	f	g	h	i	j

The start square is at c3 = ☐.

c3 + ⬌ = ___ + ___ = ____	c3 + ⤢ = __ + __ = ___
☐ + ⬌ = ___ + ___ = ___	☐ + ⤢ = __ + __ = ___
c3 + ⬌ = ___ + ___ = ____	c3 + ⤢ = __ + __ = ___
☐ + ⬌ = ___ + ___ = ____	☐ + ⤢ = __ + __ = ___

Computing additions through math and chess integrated puzzle

5			g4	
4	g3	⊤	⊤	⊤	f3
3	g1	⊤	⤡	⊤	f1
2	g2	⊢	⊤	⊤	f2
1		f4		
	a	b	c	d	e

5	⤡	5	✳	9	✳
4	✳	⤡	6	7	11
3	⤓	2	⊢	⤢	12
2	4	⊢		14	⤡
1	⬌	7	8	⤡	15
	f	g	h	i	j

The start square is at c3 = ☐.

c3 + ⬍ = ___ + ___ = ____

☐ + ⬍ = ___ + ___ = ____

c3 + ⬍ = ___ + ___ = ____

☐ + ⬍ = ___ + ___ = ____

c3 + ⤫ = __ + __ = __

☐ + ⤫ = __ + __ = __

c3 + ⤫ = __ + __ = __

☐ + ⤫ = __ + __ = __

Ho Math Chess 何数棋谜　妈！我会棋谜式加法啦！
Mom! I Learn Addition Using Math-Chess-Puzzles Connection
Contents include both traditional and Math-Chess-Puzzles combined methods. Extra strength
Student Name _____ Date _____

Computing additions through math and chess integrated puzzle

5			g4	
4	g3	⊥	⊥	⊥ f3
3	g1	⊥	※	⊥ f1
2	g2	⊥	⊥	⊥ f1
1		f4		
	a	b	c	d	e

5	⤨	5	※	9	※
4	※	⤢	6	7	11
3	↓	2	⊥	⤢	12
2	4	⊥		14	⤢
1	⚓	7	8	⤢	15
	f	g	h	i	j

The start square is at c3 = ☐ .

c3 + ⬍⬌ = ___ + ___ = ____

c3 + ⤢ = __ + __ = __

☐ + ⬍⬌ = ___ + ___ = ____

☐ + ⤢ = __ + __ = __

c3 + ⬍⬌ = ___ + ___ = ____

c3 + ⤢ = __ + __ = __

☐ + ⬍⬌ = ___ + ___ = ____

☐ + ⤢ = __ + __ = __

☐ + ⬍⬌ = ___ + ___ = ____

☐ + ⤢ = __ + __ = __

Student Name _____ Date _____

100 table and pattern

10	10	20	30	40	50	60	70	80	90	100
9	9	18	27	36	45	54	63	72	81	90
8	8	16	24	32	40	48	56	64	72	80
7	7	14	21	28	35	42	49	56	63	70
6	6	12	18	24	30	36	42	48	54	60
5	5	10	15	20	25	30	35	40	45	50
4	4	8	12	16	20	24	28	32	36	40
3	3	6	9	12	15	18	21	24	27	30
2	2	4	6	8	10	12	14	16	18	20
1	1	2	3	4	5	6	7	8	9	10
	a	b	c	d	e	f	g	h	i	j

The pattern rule of b1 to b10 is _____.

The pattern rule of c1 to c10 is _____.

Start at 3 and then count by 2.

The pattern rule of d1 to d10 is _____.

The pattern rule of e1 to e10 is _____.

The pattern rule of f1 to f10 is_____.

The pattern rule of g1 to g10 is _____.

The pattern rule of h1 to h10 is _____.

Student Name _____ Date _____

The pattern rule of i1 to i10 is _____.

The pattern rule of j1 to j10 is _____.

100 table and pattern

10	91	92	93	94	95	96	97	98	99	100
9	81	82	83	84	85	86	87	88	89	90
8	71	72	73	74	75	76	77	78	79	80
7	61	62	63	64	65	66	67	68	69	70
6	51	52	53	54	55	56	57	58	59	60
5	41	42	43	44	45	46	47	48	49	50
4	31	32	33	34	35	36	37	38	39	40
3	21	22	23	24	25	26	27	28	29	30
2	11	12	13	14	15	16	17	18	19	20
1	1	2	3	4	5	6	7	8	9	10
	a	b	c	d	e	f	g	h	i	j

Start at f7 , the pattern rule is to subtract _____ to get the next number.

Start at f7 , the pattern rule is to subtract _____ to get the next number.

Start at e1 , the pattern rule is to add _____ to get the next number.

Start at e1 , the pattern rule is to add _____ to get the next number.

Start at a5 , the pattern rule is to add _____ to get the next number.

Start at i1 , the pattern rule is to add _____ to get the next number.

100 table and pattern

10	91	92	93	94	95	96	97	98	99	100
9	81	82	83	84	85	86	87	88	89	90
8	71	72	73	74	75	76	77	78	79	80
7	61	62	63	64	65	66	67	68	69	70
6	51	52	53	54	55	56	57	58	59	60
5	41	42	43	44	45	46	47	48	49	50
4	31	32	33	34	35	36	37	38	39	40
3	21	22	23	24	25	26	27	28	29	30
2	11	12	13	14	15	16	17	18	19	20
1	1	2	3	4	5	6	7	8	9	10
	a	b	c	d	e	f	g	h	i	j

Start at a1, count every 2, you will get _____.

Start at b1, count every 2, you will get _____.

Start at g1, move along ⊥, you will get _____.

Start at e1, move along, you will get _____.

Start at d10, move along ⊥, you will get _____.

Ho Math Chess 何数棋谜　妈! 我会棋谜式加法啦!

Mom! I Learn Addition Using Math-Chess-Puzzles Connection

Contents include both traditional and Math-Chess-Puzzles combined methods. Extra strength

©2008 − 2018 Frank Ho, Amanda Ho　　All rights reserved. www.homathchess.com

Student Name _____ Date _____

Using dd + d or d + dd adding up to 10 concept using chess moves

3	5	15	35
2	25	5	45
1	65	75	55
	a	b	c

You are a chess piece and located at b2.

The original square is at b2.	Horizontal from	Vertical form	The original square is at b2.	Horizontal from	Vertical form
b2+	5 + 15 = 20	5 + 15 20	+b2	15 + 5 = 20	15 + 5 20
b2+	5 + 75 = 80	5 + 75 80	+b2	75+5=80	75 + 5 80

Using adding up to 10 concept

3	5	15	35
2	25	5	45
1	65	75	55
	a	b	c

You are a chess piece and located at b2.

The original square is at b2.	Horizontal from	Vertical form	The original square is at b2.	Horizontal from	Vertical form
b2+ ✛			✛ +b2		
b2+ ⤡			⤡ +b2		

Using adding up to 10 concept

3	5	15	35
2	25	5	45
1	65	75	55
	a	b	c

You are a chess piece and located at b2.

The original square is at b2.	Horizontal from	Vertical form	The original square is at b2.	Horizontal from	Vertical form
b2+			+b2		
b2+			+b2		

Using adding up to 10 concept

3	5	15	35	You are a chess piece and located at b2.
2	25	5	45
1	65	75	55
	a	b	c

The original square is at b2.	Horizontal from	Vertical form	The original square is at b2.	Horizontal from	Vertical form
b2+			+b2		
b2+			+b2		

Using adding up to 10 concept

				You are a chess piece and located at b2.
3	7	17	37	
2	27	3	47	
1	67	77	57	
	a	b	c	

The original square is at b2.	Horizontal from	Vertical form	The original square is at b2.	Horizontal from	Vertical form
b2+			+b2		
b2+			+b2		

Student Name _____ Date _____

Using adding up to 10 concept

					You are a chess piece and located at b2.
3	7	17	37		
2	27	3	47		
1	67	77	57		
	a	b	c		

The original square is at b2.	Horizontal from	Vertical form	The original square is at b2.	Horizontal from	Vertical form
b2+			+b2		
b2+			+b2		

Using adding up to 10 concept

3	7	17	37	You are a chess piece and located at b2.
2	27	3	47	
1	67	77	57	
	a	b	c	

The original square is at b2.	Horizontal from	Vertical form	The original square is at b2.	Horizontal from	Vertical form
b2+			+b2		
b2+			+b2		

Using adding up to 10 concept

3	7	17	37	You are a chess piece and located at b2.
2	27	3	47	
1	67	77	57	
	a	b	c	

The original square is at b2.	Horizontal from	Vertical form	The original square is at b2.	Horizontal from	Vertical form
b2+			+b2		
b2+			+b2		

Using adding up to 10 concept

3	6	16	36
2	26	4	46
1	66	76	56
	a	b	c

You are a chess piece and located at b2.

The original square is at b2.	Horizontal from	Vertical form	The original square is at b2.	Horizontal from	Vertical form
b2+			+b2		
b2+			+b2		

Using adding up to 10 concept

3	6	16	36	You are a chess piece and located at b2.
2	26	4	46	
1	66	76	56	
	a	b	c	

The original square is at b2.	Horizontal from	Vertical form	The original square is at b2.	Horizontal from	Vertical form
b2+ ⬍⬌			⬍⬌ +b2		
b2+ ⬍⬌			⬍⬌ +b2		

Using adding up to 10 concept

3	6	16	36	You are a chess piece and located at b2.
2	26	4	46	
1	66	76	56	
	a	b	c	

The original square is at b2.	Horizontal from	Vertical form	The original square is at b2.	Horizontal from	Vertical form
b2+			+b2		
b2+			+b2		

Student Name _____ Date _____

Using adding up to 10 concept

3	6	16	36	You are a chess piece and located at b2.
2	26	4	46	
1	66	76	56	
	a	b	c	

The original square is at b2.	Horizontal from	Vertical form	The original square is at b2.	Horizontal from	Vertical form
b2+			+b2		
b2+			+b2		

Student Name _____ Date _____

Adding up to 10

3	6	16	36	You are a chess piece and located at b2.
2	26	4	46	
1	66	76	56	
	a	b	c	

The original square is at b2.	Horizontal from	Vertical form	The original square is at b2.	Horizontal from	Vertical form
b2+			+b2		
b2+			+b2		

Student Name _____　　Date _____

Using adding up to 10 concept

3	8	18	38		You are a chess piece and located at b2.
2	28	2	48		
1	68	78	58		
	a	b	c		

The original square is at b2.	Horizontal from	Vertical form	The original square is at b2.	Horizontal from	Vertical form
b2+ ⬌⬍			⬌⬍ +b2		
b2+ ⬌⬍			⬌⬍ +b2		

Using adding up to 10 concept

3	8	18	38		You are a chess piece and located at b2.
2	28	2	48		
1	68	78	58		
	a	b	c		

The original square is at b2.	Horizontal from	Vertical form	The original square is at b2.	Horizontal from	Vertical form
b2+		2 + 8 10	+b2		8 + 2 10
b2+		2 + 58 60	+b2		58 + 2 60

Using adding up to 10 concept

3	8	18	38	You are a chess piece and located at b2.
2	28	2	48	
1	68	78	58	
	a	b	c	

The original square is at b2.	Horizontal from	Vertical form	The original square is at b2.	Horizontal from	Vertical form
b2+ ✕	40		✕ +b2		
b2+ ✕	70		✕ +b2		

Using adding up to 10 concept

3	9	19	39	You are a chess piece and located at b2.
2	29	1	49	
1	66	79	59	
	a	b	c	

The original square is at b2.	Horizontal from	Vertical form	The original square is at b2.	Horizontal from	Vertical form
b2+	20		+b2		
b2+	80		+b2		

Using adding up to 10 concept

3	9	19	39		You are a chess piece and located at b2.
2	29	1	49		
1	66	79	59		
	a	b	c		

The original square is at b2.	Horizontal from	Vertical form	The original square is at b2.	Horizontal from	Vertical form
b2+	30		+b2		
b2+	50		+b2		

Using adding up to 10 concept

3	9	19	39
2	29	1	49
1	66	79	59
	a	b	c

You are a chess piece and located at b2.

The original square is at b2.	Horizontal from	Vertical form	The original square is at b2.	Horizontal from	Vertical form
b2+✕	10		✕+b2		
b2+✕	60		✕+b2		

Mom! I Learn Addition Using Math-Chess-Puzzles Connection

Student Name _____ Date _____

Using adding up to 10 concept

3	9	19	39
2	29	1	49
1	66	79	59
	a	b	c

You are a chess piece and located at b2.

The original square is at b2.	Horizontal from	Vertical form	The original square is at b2.	Horizontal from	Vertical form
b2+	40		+b2		
b2+	67		+b2		

Addition using chess moves

e	2	2	3	4	7
d	6	4	8	9	0
c	1	2	2	3	4
b	5	6	7	6	9
a	0	1	2	3	8
	1	2	3	4	5

Move by one square at a time from the original square.

The original square is at c3.

C3 $+$ ⬌ $=$ _____ $+$ _____ $=$ _____ 2 $+$ 3 $=$ 5

The original square is at c3.

C3 $+$ ⬌ $=$ _____ $+$ _____ $=$ _____

The original square is at c3.

C3 $+$ ⬌ $=$ _____ $+$ _____ $=$ _____

The original square is at c3.

C3 $+$ ⬌ $=$ _____ $+$ _____ $=$ _____

Student Name _____ Date _____

Addition

e	1	2	3	4	5
d	6	7	8	9	0
c	1	2	5	3	4
b	5	6	7	8	9
a	0	1	2	3	4
	1	2	3	4	5

Move by one square at a time from the original square.

The original square is at c3.

c3 $+$ $+$ $=$ _____ $+$ _____ $+$ _____ $=$ _____

The original square is at c3.

c3 $+$ $+$ $=$ _____ $+$ _____ $+$ _____ $=$ _____

The original square is at c3.

c3 $+$ $+$ $=$ _____ $+$ _____ $+$ _____ $=$ _____

The original square is at c3.

c3 $+$ $+$ $=$ _____ $+$ _____ $+$ _____ $=$ _____

Student Name _____ Date _____

Addition

e	1	2	3	4	5
d	6	7	8	19	2
🍦	1	2	7	3	4
b	5	6	7	8	9
a	3	3	3	3	4
	1	2	3	4	5

Move by one square at a time from the original square.

The original square is at c3.

🍦₃ + ⬌⬍ + ⬌⬍ +

= __7__ + __3__ + __4__ = _____

The original square is at c3.

c3 + ⬌⬍ + ⬌⬍ = ____ + ____ + _____ = _7+8+3=_____

The original square is at c3.

c3 + ⬌⬍ + ⬌⬍ = ____ + ____ + _____ = _____

The original square is at c3.

c3 + ⬌⬍ + ⬌⬍ = ____ + ____ + _____ = _____

Addition

e	11	12	13	14	15
d	16	17	18	19	20
c	21	22	2	23	24
b	25	26	27	28	29
a	30	31	32	33	34
	1	2	3	4	5

Move by one square at a time from the original square.

The original square is at c3.

c3 $+$ ⤡⤢ $=$ _____ $+$ _____ $=$ _____

The original square is at c3.

c3 $+$ ⤡⤢ $=$ _____ $+$ _____ $=$ _____

The original square is at c3.

c3 $+$ ⤡⤢ $=$ _____ $+$ _____ $=$ _____

The original square is at c3.

c3 $+$ ⤡⤢ $=$ _____ $+$ _____ $=$ _____

Addition

e	11	12	13	14	15
d	16	17	18	19	20
c	21	22	2	23	24
b	25	26	27	28	29
a	30	31	32	33	34
	1	2	3	4	5

Move by one square at a time from the original square.

The original square is at c3.

C3 ＋ ⬌ = _____ ＋ _____ = _____

The original square is at c3.

C3 ＋ ⬍ = _____ ＋ _____ = _____

The original square is at c3.

C3 ＋ ⬍ = _____ ＋ _____ = _____

The original square is at c3.

C3 ＋ ⬍ = _____ ＋ _____ = _____

Student Name _____ Date _____

Addition

e	11	12	13	14	15
d	16	17	18	19	20
c	21	22	2	23	24
b	25	26	27	28	29
a	30	31	32	33	34
	1	2	3	4	5

Move by one square at a time from where you are.

The original square is at c3.

c3 $+$ $+$ $=$ ____ $+$ ____ $+$ ____ $=$ _____

The original square is at c3.

c3 $+$ $+$ $=$ ____ $+$ ____ $+$ ____ $=$ _____

The original square is at c3.

c3 $+$ $+$ $=$ ____ $+$ ____ $+$ ____ $=$ _____

The original square is at c3.

c3 $+$ $+$ $=$ ____ $+$ ____ $+$ ____ $=$ _____

Student Name _____ Date _____

Addition

e	11	12	13	14	15
d	16	17	18	19	20
🍦	21	22	2	23	24
b	25	26	27	28	29
a	30	31	32	33	34
	1	2	3	4	5

Move by one square at a time from where you are.

The original square is at c3.

🍦₃ + ⬌⬍ + ⬌⬍

= _____ + _____ + _____ = _____

The original square is at c3.

c3 + ⬌⬍ + ⬌⬍ = _____ + _____ + _____ = _____

The original square is at c3.

c3 + ⬌⬍ + ⬌⬍ = _____ + _____ + _____ = _____

The original square is at c3.

c3 + ⬌⬍ + ⬌⬍ = _____ + _____ + _____ = _____

Addition

e	11	12	13	14	15
d	16	17	18	19	20
c	21	22	2	23	24
b	25	26	27	28	29
a	30	31	32	33	34
	1	2	3	4	5

Move by one square at a time from the original square.

The original square is at c3.

c3 $+ \times =$ _____ $+$ _____ $=$ _____

The original square is at c3.

c3 $+ \times =$ _____ $+$ _____ $=$ _____

The original square is at c3.

c3 $+ \times =$ _____ $+$ _____ $=$ _____

The original square is at c3.

c3 $+ \times =$ _____ $+$ _____ $=$ _____

Student Name _____　Date _____

Addition

e	11	12	13	14	15
d	16	17	18	19	20
c	21	22	2	23	24
b	25	26	27	28	29
a	30	31	32	33	34
	1	2	3	4	5

Move by one square at a time from the original square.

The original square is at c3.

C3 $+$ $+$ $=$ _____ $+$ _____ $+$ _____ $=$ _____

The original square is at c3.

C3 $+$ $+$ $=$ _____ $+$ _____ $+$ _____ $=$ _____

The original square is at c3.

C3 $+$ $+$ $=$ _____ $+$ _____ $+$ _____ $=$ _____

The original square is at c3.

C3 $+$ $+$ $=$ _____ $+$ _____ $+$ _____ $=$ _____

Student Name _____ Date _____

Addition

e	11	12	13	14	15
d	16	17	18	19	20
c	21	22	2	23	24
b	25	26	27	28	29
a	30	31	32	33	34
	1	2	3	4	5

Move by one square at a time from the original square.

The original square is at c3.

C3 $+ \times + \times$ = _____ + _____ + _____ = _____

The original square is at c3.

C3 $+ \times + \times$ = _____ + _____ + _____ = _____

The original square is at c3.

C3 $+ \times + \times$ = _____ + _____ + _____ = _____

The original square is at c3.

C3 $+ \times + \times$ = _____ + _____ + _____ = _____

Student Name _____ Date _____

Addition

e	11	12	13	14	15
d	16	17	18	19	20
c	21	22	2	23	24
b	25	26	27	28	29
a	30	31	32	33	34
	1	2	3	4	5

Move by one square at a time from the original square.

The original square is at c3.

C3 ✛ ⊥ = _____ + _____ = _____

The original square is at c3.

C3 ✛ ⊥ = _____ + _____ = _____

The original square is at c3.

C3 ✛ ⊥ = _____ + _____ = _____

The original square is at c3.

C3 ✛ ⊥ = _____ + _____ = _____

Student Name _____ Date _____

Addition

e	11	12	13	14	15
d	16	17	18	19	20
c	21	22	2	23	24
b	25	26	27	28	29
a	30	31	32	33	34
	1	2	3	4	5

Move by one square at a time from the original square.

The original square is at c3.

C3 + ┤┬├ + ┤┬├ = ____ + ____ + ____ = ____

The original square is at c3.

C3 + ┤┬├ + ┤┬├ = ____ + ____ + ____ = ____

The original square is at c3.

C3 + ┤┬├ + ┤┬├ = ____ + ____ + ____ = ____

The original square is at c3.

C3 + ┤┬├ + ┤┬├ = ____ + ____ + ____ = ____

Student Name _____ Date _____

Addition

e	11	12	13	14	15
d	16	17	18	19	20
c	21	22	2	23	24
b	25	26	27	28	29
a	30	31	32	33	34
	1	2	3	4	5

Move by one square at a time from the original square.

The original square is at c3.

C3 $+$ = _____ $+$ _____ = _____

The original square is at c3.

C3 $+$ = _____ $+$ _____ = _____

The original square is at c3.

C3 $+$ = _____ $+$ _____ = _____

The original square is at c3.

C3 $+$ = _____ $+$ _____ = _____

Addition

e	11	12	13	14	15
d	16	17	18	19	20
c	21	22	2	23	24
b	25	26	27	28	29
a	30	31	32	33	34
	1	2	3	4	5

Move by one square at a time from the original square.

The original square is at c3.

C3 = _____ + _____ + _____ = _____

The original square is at c3.

C3 = _____ + _____ + _____ = _____

The original square is at c3.

C3 = _____ + _____ + _____ = _____

The original square is at c3.

C3 = _____ + _____ + _____ = _____

Addition

e	11	12	13	14	15
d	16	17	18	19	20
c	21	22	3	23	24
b	25	26	27	28	29
a	30	31	32	33	34
	1	2	3	4	5

Move by one square at a time from the original square.

The original square is at c3.

C3 $+$ = _____ $+$ _____ = _____

The original square is at c3.

C3 $+$ = _____ $+$ _____ = _____

The original square is at c3.

C3 $+$ = _____ $+$ _____ = _____

The original square is at c3.

C3 $+$ = _____ $+$ _____ = _____

Addition

e	11	12	13	14	15
d	16	17	18	19	20
c	21	22	3	23	24
b	25	26	27	28	29
a	30	31	32	33	34
	1	2	3	4	5

Move by one square at a time from the original square.

The original square is at c3.

C3 + + = ____ + ____ + ____ = _____

The original square is at c3.

C3 + + = ____ + ____ + ____ = _____

The original square is at c3.

C3 + + = ____ + ____ + ____ = _____

The original square is at c3.

C3 + + = ____ + ____ + ____ = _____

Addition

e	11	12	13	14	15
d	16	17	18	19	20
c	21	22	3	23	24
b	25	26	27	28	29
a	30	31	32	33	34
	1	2	3	4	5

Move by one square at a time from the original square.

The original square is at c3.

C3 $+$ $+$ $=$ ____ $+$ ____ $+$ ____ $=$ ____

The original square is at c3.

C3 $+$ $+$ $=$ ____ $+$ ____ $=$ ____

The original square is at c3.

C3 $+$ $+$ $=$ ____ $+$ ____ $=$ ____

The original square is at c3.

C3 $+$ $+$ $=$ ____ $+$ ____ $=$ ____

Student Name _____ Date _____

Addition

e	11	12	13	14	15
d	16	17	18	19	20
c	21	22	3	23	24
b	25	26	27	28	29
a	30	31	32	33	34
	1	2	3	4	5

Move by one square at a time from the original square.

The original square is at c3.

c3 + ✕ = _____ + _____ = _____

The original square is at c3.

c3 + ✕ = _____ + _____ = _____

The original square is at c3.

c3 + ✕ = _____ + _____ = _____

The original square is at c3.

c3 + ✕ = _____ + _____ = _____

Addition

e	11	12	13	14	15
d	16	17	18	19	20
c	21	22	3	23	24
b	25	26	27	28	29
a	30	31	32	33	34
	1	2	3	4	5

Move by one square at a time from the original square.

The original square is at c3.

c3 $+ \times + \times =$ ____ + ____ + ____ = ____

The original square is at c3.

c3 $+ \times + \times =$ ____ + ____ + ____ = ____

The original square is at c3.

c3 $+ \times + \times =$ ____ + ____ + ____ = ____

The original square is at c3.

c3 $+ \times + \times =$ ____ + ____ + ____ = ____

Student Name _____ Date _____

Addition

e	11	12	13	14	15
d	16	17	18	19	20
c	21	22	3	23	24
b	25	26	27	28	29
a	30	31	32	33	34
	1	2	3	4	5

Move by one square at a time from the original square.

The original square is at c3.

c3 $+$ $+$ $=$ _____ $+$ _____ $+$ _____ $=$ _____

The original square is at c3.

c3 $+$ $+$ $=$ _____ $+$ _____ $+$ _____ $=$ _____

The original square is at c3.

c3 $+$ $+$ $=$ _____ $+$ _____ $+$ _____ $=$ _____

The original square is at c3.

c3 $+$ $+$ $=$ _____ $+$ _____ $+$ _____ $=$ _____

Student Name _____ Date _____

Addition

e	11	12	13	14	15
d	16	17	18	19	20
c	21	22	3	23	24
b	25	26	27	28	29
a	30	31	32	33	34
	1	2	3	4	5

Move by one square at a time from the original square.

The original square is at c3.

C3 ➕ ⊥ = _____ ➕ _____ = _____

The original square is at c3.

C3 ➕ ⊥ = _____ ➕ _____ = _____

The original square is at c3.

C3 ➕ ⊥ = _____ ➕ _____ = _____

The original square is at c3.

C3 ➕ ⊥ = _____ ➕ _____ = _____

Addition

e	11	12	13	14	15
d	16	17	18	19	20
c	21	22	3	23	24
b	25	26	27	28	29
a	30	31	32	33	34
	1	2	3	4	5

Move by one square at a time from the original square.

The original square is at c3.

C3 $+$ ┼ = _____ $+$ _____ = _____

The original square is at c3.

C3 $+$ ┼ = _____ $+$ _____ = _____

The original square is at c3.

C3 $+$ ┼ = _____ $+$ _____ = _____

The original square is at c3.

C3 $+$ ┼ = _____ $+$ _____ = _____

Addition

e	11	12	13	14	15
d	16	17	18	19	20
c	21	22	4	23	24
b	25	26	27	28	29
a	30	31	32	33	34
	1	2	3	4	5

Move by one square at a time from the original square.

The original square is at c3.

C3 $+$ ⬌ $=$ _____ $+$ _____ $=$ _____

The original square is at c3.

c3 $+$ ⬌ $=$ _____ $+$ _____ $=$ _____

The original square is at c3.

C3 $+$ ⬌ $=$ _____ $+$ _____ $=$ _____

The original square is at c3.

C3 $+$ ⬌ $=$ _____ $+$ _____ $=$ _____

Student Name _____ Date _____

Addition

e	11	12	13	14	15
d	16	17	18	19	20
c	21	22	2	23	24
b	25	26	27	28	29
a	30	31	32	33	34
	1	2	3	4	5

Move by one square at a time from the original square.

The original square is at c3.

C3 $+$ ⬌⬍ $+$ ⬌⬍ = _____ $+$ _____ $+$ _____ = _____

The original square is at c3.

C3 $+$ ⬌⬍ $+$ ⬌⬍ = _____ $+$ _____ $+$ _____ = _____

The original square is at c3.

C3 $+$ ⬌⬍ $+$ ⬌⬍ = _____ $+$ _____ $+$ _____ = _____

The original square is at c3.

C3 $+$ ⬌⬍ $+$ ⬌⬍ = _____ $+$ _____ $+$ _____ = _____

Addition

e	11	12	13	14	15
d	16	17	18	19	20
c	21	22	4	23	24
b	25	26	27	28	29
a	30	31	32	33	34
	1	2	3	4	5

Move by one square at a time from the original square.

The original square is at c3.

C3 + + = _____ + _____ + _____ = _____

The original square is at c3.

C3 + + = _____ + _____ + _____ = _____

The original square is at c3.

C3 + + = _____ + _____ + _____ = _____

The original square is at c3.

C3 + + = _____ + _____ + _____ = _____

Student Name _____ Date _____

Addition

e	11	12	13	14	15
d	16	17	18	19	20
c	21	22	4	23	24
b	25	26	27	28	29
a	30	31	32	33	34
	1	2	3	4	5

Move by one square at a time from the original square.

The original square is at c3.

C3 $+$ ✕ = _____ $+$ _____ = _____

The original square is at c3.

C3 $+$ ✕ = _____ $+$ _____ = _____

The original square is at c3.

C3 $+$ ✕ = _____ $+$ _____ = _____

The original square is at c3.

C3 $+$ ✕ = _____ $+$ _____ = _____

Student Name _____ Date _____

Addition

e	11	12	13	14	15
d	16	17	18	19	20
c	21	22	4	23	24
b	25	26	27	28	29
a	30	31	32	33	34
	1	2	3	4	5

Move by one square at a time from the original square.

The original square is at c3.

c3 + ✕ + ✕ = _____ + _____ + _____ = _____

The original square is at c3.

c3 + ✕ + ✕ = _____ + _____ + _____ = _____

The original square is at c3.

c3 + ✕ + ✕ = _____ + _____ + _____ = _____

The original square is at c3.

c3 + ✕ + ✕ = _____ + _____ + _____ = _____

Student Name _____ Date _____

Addition

e	11	12	13	14	15
d	16	17	18	19	20
c	21	22	4	23	24
b	25	26	27	28	29
a	30	31	32	33	34
	1	2	3	4	5

Move by one square at a time from the original square.

The original square is at c3.

c3 + ✕ + ✕ = _____ + _____ + _____ = _____

The original square is at c3.

c3 + ✕ + ✕ = _____ + _____ + _____ = _____

The original square is at c3.

c3 + ✕ + ✕ = _____ + _____ + _____ = _____

The original square is at c3.

c3 + ✕ + ✕ = _____ + _____ + _____ = _____

Addition

e	11	12	13	14	15
d	16	17	18	19	20
c	21	22	4	23	24
b	25	26	27	28	29
a	30	31	32	33	34
	1	2	3	4	5

Move by one square at a time from the original square.

The original square is at c3.

C3 ✛ ⊥ = _____ ✛ _____ = _____

The original square is at c3.

C3 ✛ ⊥ = _____ ✛ _____ = _____

The original square is at c3.

C3 ✛ ⊥ = _____ ✛ _____ = _____

The original square is at c3.

C3 ✛ ⊥ = _____ ✛ _____ = _____

Addition

e	11	12	13	14	15
d	16	17	18	19	20
c	21	22	4	23	24
b	25	26	27	28	29
a	30	31	32	33	34
	1	2	3	4	5

Move by one square at a time from the original square.

The original square is at c3.

c3 $+$ ⊥ $=$ _____ $+$ _____ $=$ _____

The original square is at c3.

c3 $+$ ⊥ $=$ _____ $+$ _____ $=$ _____

The original square is at c3.

c3 $+$ ⊥ $=$ _____ $+$ _____ $=$ _____

The original square is at c3.

c3 $+$ ⊥ $=$ _____ $+$ _____ $=$ _____

440

Addition

e	11	12	13	14	15
d	16	17	18	19	20
c	21	22	5	23	24
b	25	26	27	28	29
a	30	31	32	33	34
	1	2	3	4	5

Move by one square at a time from the original square.

The original square is at c3.

c3 $+$ = _____ $+$ _____ = _____

The original square is at c3.

c3 $+$ = _____ $+$ _____ = _____

The original square is at c3.

c3 $+$ = _____ $+$ _____ = _____

The original square is at c3.

c3 $+$ = _____ $+$ _____ = _____

Student Name _____ Date _____

Addition

e	11	12	13	14	15
d	16	17	18	19	20
c	21	22	5	23	24
b	25	26	27	28	29
a	30	31	32	33	34
	1	2	3	4	5

Move by one square at a time from the original square.

The original square is at c3.

c3 $+$ ✕ = _____ $+$ _____ = _____

The original square is at c3.

c3 $+$ ✕ = _____ $+$ _____ = _____

The original square is at c3.

c3 $+$ ✕ = _____ $+$ _____ = _____

The original square is at c3.

c3 $+$ ✕ = _____ $+$ _____ = _____

Addition

e	11	12	13	14	15
d	16	17	18	19	20
c	21	22	5	23	24
b	25	26	27	28	29
a	30	31	32	33	34
	1	2	3	4	5

Move by one square at a time from the original square.

The original square is at c3.

C3 + ⤬ + ⤬ = _____ + _____ + _____ = _____

The original square is at c3.

C3 + ⤬ + ⤬ = _____ + _____ + _____ = _____

The original square is at c3.

C3 + ⤬ + ⤬ = _____ + _____ + _____ = _____

The original square is at c3.

C3 + ⤬ + ⤬ = _____ + _____ + _____ = _____

Student Name _____ Date _____

Addition

e	11	12	13	14	15
d	16	17	18	19	20
c	21	22	5	23	24
b	25	26	27	28	29
a	30	31	32	33	34
	1	2	3	4	5

Move by one square at a time from the original square.

The original square is at c3.

c3 $+ \times + \times =$ ____ $+$ ____ $+$ ____ $=$ _____

The original square is at c3.

c3 $+ \times + \times =$ ____ $+$ ____ $+$ ____ $=$ _____

The original square is at c3.

C3 $+ \times + \times =$ ____ $+$ ____ $+$ ____ $=$ _____

The original square is at c3.

c3 $+ \times + \times =$ ____ $+$ ____ $+$ ____ $=$ _____

Addition

e	11	12	13	14	15
d	16	17	18	19	20
c	21	22	5	23	24
b	25	26	27	28	29
a	30	31	32	33	34
	1	2	3	4	5

Move by one square at a time from the original square.

The original square is at c3.

C3 + ⊥ = _____ + _____ = _____

The original square is at c3.

C3 + ⊥ = _____ + _____ = _____

The original square is at c3.

C3 + ⊥ = _____ + _____ = _____

The original square is at c3.

C3 + ⊥ = _____ + _____ = _____

Addition

e	11	12	13	14	15
d	16	17	18	19	20
c	21	22	5	23	24
b	25	26	27	28	29
a	30	31	32	33	34
	1	2	3	4	5

Move by one square at a time from the original square.

The original square is at c3.

C3 + = _____ + _____ = _____

The original square is at c3.

C3 + = _____ + _____ = _____

The original square is at c3.

C3 + = _____ + _____ = _____

The original square is at c3.

C3 + = _____ + _____ = _____

Student Name _____ Date _____

Addition

e	11	12	13	14	15
d	16	17	18	19	20
c	21	22	6	23	24
b	25	26	27	28	29
a	30	31	32	33	34
	1	2	3	4	5

Move by one square at a time from the original square.

The original square is at c3.

C3 + = _____ + _____ = _____

The original square is at c3.

C3 + = _____ + _____ = _____

The original square is at c3.

C3 + = _____ + _____ = _____

The original square is at c3.

C3 + = _____ + _____ = _____

Addition

e	11	12	13	14	15
d	16	17	18	19	20
c	21	22	6	23	24
b	25	26	27	28	29
a	30	31	32	33	34
	1	2	3	4	5

Move by one square at a time from the original square.

The original square is at c3.

c3 $+$ ⬌⬍ $+$ ⬌⬍ $=$ ____ $+$ ____ $+$ _____ $=$ _____

The original square is at c3.

c3 $+$ ⬌⬍ $+$ ⬍ $=$ ____ $+$ ____ $+$ _____ $=$ _____

The original square is at c3.

c3 $+$ ⬌⬍ $+$ ⬌⬍ $=$ ____ $+$ ____ $+$ _____ $=$ _____

The original square is at c3.

c3 $+$ ⬌⬍ $+$ ⬍ $=$ ____ $+$ ____ $+$ _____ $=$ _____

Ho Math Chess 何数棋谜   妈！我会棋谜式加法啦！
Mom! I Learn Addition Using Math-Chess-Puzzles Connection
Contents include both traditional and Math-Chess-Puzzles combined methods. Extra strength
©2008 – 2018 Frank Ho, Amanda Ho   All rights reserved. www.homathchess.com

Student Name _____ Date _____

Addition

e	11	12	13	14	15
d	16	17	18	19	20
c	21	22	6	23	24
b	25	26	27	28	29
a	30	31	32	33	34
	1	2	3	4	5

Move by one square at a time from the original square.

The original square is at c3.

C3 + + = ____ + ____ + ____ = _____

The original square is at c3.

C3 + + = ____ + ____ + ____ = _____

The original square is at c3.

C3 + + = ____ + ____ + ____ = _____

The original square is at c3.

C3 + + = ____ + ____ + ____ = _____

Ho Math Chess 何数棋谜   妈！我会棋谜式加法啦！
Mom! I Learn Addition Using Math-Chess-Puzzles Connection
Contents include both traditional and Math-Chess-Puzzles combined methods. Extra strength

©2008 – 2018 Frank Ho, Amanda Ho   All rights reserved. www.homathchess.com

Student Name _____ Date _____

Addition

e	11	12	13	14	15
d	16	17	18	19	20
c	21	22	6	23	24
b	25	26	27	28	29
a	30	31	32	33	34
	1	2	3	4	5

Move by one square at a time from the original square.

The original square is at c3.

C3 + ✕ = _____ + _____ = _____

The original square is at c3.

C3 + ✕ = _____ + _____ = _____

The original square is at c3.

C3 + ✕ = _____ + _____ = _____

The original square is at c3.

C3 + ✕ = _____ + _____ = _____

Ho Math Chess 何数棋谜　妈！我会棋谜式加法啦！

Mom! I Learn Addition Using Math-Chess-Puzzles Connection

Contents include both traditional and Math-Chess-Puzzles combined methods. Extra strength

©2008 – 2018 Frank Ho, Amanda Ho　All rights reserved. www.homathchess.com

Student Name _____ Date _____

Addition

e	11	12	13	14	15
d	16	17	18	19	20
c	21	22	6	23	24
b	25	26	27	28	29
a	30	31	32	33	34
	1	2	3	4	5

Move by one square at a time from the original square.

The original square is at c3.

c3 $+ \times + \times =$ ____ + ____ + ____ = _____

The original square is at c3.

c3 $+ \times + \times =$ ____ + ____ + ____ = _____

The original square is at c3.

c3 $+ \times + \times =$ ____ + ____ + ____ = _____

The original square is at c3.

c3 $+ \times + \times =$ ____ + ____ + ____ = _____

Addition

e	11	12	13	14	15
d	16	17	18	19	20
c	21	22	6	23	24
b	25	26	27	28	29
a	30	31	32	33	34
	1	2	3	4	5

Move by one square at a time from the original square.

The original square is at c3.

c3 $+$ ⤡⤢ $+$ ⤢⤡ = ____ $+$ ____ $+$ ____ = _____

The original square is at c3.

c3 $+$ ⤢⤡ $+$ ⤡⤢ = ____ $+$ ____ $+$ ____ = _____

The original square is at c3.

c3 $+$ ⤡⤢ $+$ ⤢⤡ = ____ $+$ ____ $+$ ____ = _____

The original square is at c3.

c3 $+$ ⤢⤡ $+$ ⤡⤢ = ____ $+$ ____ $+$ ____ = _____

Student Name _____ Date _____

Addition

e	11	12	13	14	15
d	16	17	18	19	20
c	21	22	6	23	24
b	25	26	27	28	29
a	30	31	32	33	34
	1	2	3	4	5

Move by one square at a time from the original square.

The original square is at c3.

C3 ➕ ┿ = _____ ➕ _____ = _____

The original square is at c3.

C3 ➕ ┿ = _____ ➕ _____ = _____

The original square is at c3.

C3 ➕ ┿ = _____ ➕ _____ = _____

The original square is at c3.

C3 ➕ ┿ = _____ ➕ _____ = _____

Ho Math Chess 何数棋谜　妈！我会棋谜式加法啦！

Mom! I Learn Addition Using Math-Chess-Puzzles Connection

Contents include both traditional and Math-Chess-Puzzles combined methods. Extra strength

©2008 – 2018 Frank Ho, Amanda Ho　All rights reserved. www.homathchess.com

Student Name _____ Date _____

Addition

e	11	12	13	14	15
d	16	17	18	19	20
c	21	22	6	23	24
b	25	26	27	28	29
a	30	31	32	33	34
	1	2	3	4	5

Move by one square at a time from the original square.

The original square is at c3.

C3 $+$ ⊥ $=$ _____ $+$ _____ $=$ _____

The original square is at c3.

C3 $+$ ⊥ $=$ _____ $+$ _____ $=$ _____

The original square is at c3.

C3 $+$ ⊥ $=$ _____ $+$ _____ $=$ _____

The original square is at c3.

C3 $+$ ⊥ $=$ _____ $+$ _____ $=$ _____

Student Name _____ Date _____

Addition

e	11	12	13	14	15
d	16	17	18	19	20
c	21	22	7	23	24
b	25	26	27	28	29
a	30	31	32	33	34
	1	2	3	4	5

Move by one square at a time from the original square.

The original square is at c3.

C3 $+$ = _____ $+$ _____ = _____

The original square is at c3.

C3 $+$ = _____ $+$ _____ = _____

The original square is at c3.

C3 $+$ = _____ $+$ _____ = _____

The original square is at c3.

C3 $+$ = _____ $+$ _____ = _____

Addition

e	11	12	13	14	15
d	16	17	18	19	20
c	21	22	7	23	24
b	25	26	27	28	29
a	30	31	32	33	34
	1	2	3	4	5

Move by one square at a time from the original square.

The original square is at c3.

c3 $+ \times =$ _____ $+$ _____ $=$ _____26

The original square is at c3.

c3 $+ \times =$ _____ $+$ _____ $=$ _____

The original square is at c3.

C3 $+ \times =$ _____ $+$ _____ $=$ _____

The original square is at c3.

C3 $+ \times =$ _____ $+$ _____ $=$ _____

Student Name _____ Date _____

Addition

e	11	12	13	14	15
d	16	17	18	19	20
c	21	22	7	23	24
b	25	26	27	28	29
a	30	31	32	33	34
	1	2	3	4	5

Move by one square at a time from the original square.

The original square is at c3.

$c3 + $ $ + $ $ = $ ____ $ + $ ____ $ + $ ____ $ = $ _____

The original square is at c3.

$c3 + $ $ + $ $ = $ ____ $ + $ ____ $ + $ ____ $ = $ _____

The original square is at c3.

$c3 + $ $ + $ $ = $ ____ $ + $ ____ $ + $ ____ $ = $ _____

The original square is at c3.

$c3 + $ $ + $ $ = $ ____ $ + $ ____ $ + $ ____ $ = $ _____

Addition

e	11	12	13	14	15
d	16	17	18	19	20
c	21	22	7	23	24
b	25	26	27	28	29
a	30	31	32	33	34
	1	2	3	4	5

Move by one square at a time from the original square.

The original square is at c3.

c3 $+$ $+$ $=$ ____ $+$ ____ $+$ ____ $=$ _____

The original square is at c3.

c3 $+$ $+$ $=$ ____ $+$ ____ $+$ ____ $=$ _____

The original square is at c3.

c3 $+$ $+$ $=$ ____ $+$ ____ $+$ ____ $=$ _____

The original square is at c3.

c3 $+$ $+$ $=$ ____ $+$ ____ $+$ ____ $=$ _____

Addition

e	11	12	13	14	15
d	16	17	18	19	20
c	21	22	7	23	24
b	25	26	27	28	29
a	30	31	32	33	34
	1	2	3	4	5

Move by one square at a time from the original square.

The original square is at c3.

C3 $+$ $\vdash\!\!\top\!\!\dashv$ $=$ _____ $+$ _____ $=$ _____

The original square is at c3.

C3 $+$ $\vdash\!\!\top\!\!\dashv$ $=$ _____ $+$ _____ $=$ _____

The original square is at c3.

C3 $+$ $\vdash\!\!\top\!\!\dashv$ $=$ _____ $+$ _____ $=$ _____

The original square is at c3.

C3 $+$ $\vdash\!\!\top\!\!\dashv$ $=$ _____ $+$ _____ $=$ _____

Addition

e	11	12	13	14	15
d	16	17	18	19	20
c	21	22	7	23	24
b	25	26	27	28	29
a	30	31	32	33	34
	1	2	3	4	5

Move by one square at a time from the original square.

The original square is at c3.

C3 + ┼ = _____ + _____ = _____

The original square is at c3.

C3 + ┼ = _____ + _____ = _____

The original square is at c3.

C3 + ┼ = _____ + _____ = _____

The original square is at c3.

C3 + ┼ = _____ + _____ = _____

Addition

e	11	12	13	14	15
d	16	17	18	19	20
c	21	22	8	23	24
b	25	26	27	28	29
a	30	31	32	33	34
	1	2	3	4	5

Move by one square at a time from the original square.

The original square is at c3.

C3 $+$ $=$ _____ $+$ _____ $=$ _____

The original square is at c3.

C3 $+$ $=$ _____ $+$ _____ $=$ _____

The original square is at c3.

C3 $+$ $=$ _____ $+$ _____ $=$ _____

The original square is at c3.

C3 $+$ $=$ _____ $+$ _____ $=$ _____

Addition

e	11	12	13	14	15
d	16	17	18	19	20
c	21	22	8	23	24
b	25	26	27	28	29
a	30	31	32	33	34
	1	2	3	4	5

Move by one square at a time from the original square.

The original square is at c3.

c3 $+$ ⤢ = _____ $+$ _____ = _____

The original square is at c3.

c3 $+$ ⤢ = _____ $+$ _____ = _____

The original square is at c3.

c3 $+$ ⤢ = _____ $+$ _____ = _____

The original square is at c3.

c3 $+$ ⤢ = _____ $+$ _____ = _____

Student Name _____ Date _____

Addition

e	11	12	13	14	15
d	16	17	18	19	20
c	21	22	8	23	24
b	25	26	27	28	29
a	30	31	32	33	34
	1	2	3	4	5

Move by one square at a time from the original square.

The original square is at c3.

C3 $+$ ⤬ $+$ ⤬ = ____ $+$ ____ $+$ ____ = _____

The original square is at c3.

C3 $+$ ⤬ $+$ ⤬ = ____ $+$ ____ $+$ ____ = _____

The original square is at c3.

C3 $+$ ⤬ $+$ ⤬ = ____ $+$ ____ $+$ ____ = _____

The original square is at c3.

C3 $+$ ⤬ $+$ ⤬ = ____ $+$ ____ $+$ ____ = _____

Addition

e	11	12	13	14	15
d	16	17	18	19	20
c	21	22	8	23	24
b	25	26	27	28	29
a	30	31	32	33	34
	1	2	3	4	5

Move by one square at a time from the original square.

The original square is at c3.

c3 $+$ $+$ $=$ ____ $+$ ____ $+$ ____ $=$ _____

The original square is at c3.

c3 $+$ $+$ $=$ ____ $+$ ____ $+$ ____ $=$ _____

The original square is at c3.

c3 $+$ $+$ $=$ ____ $+$ ____ $+$ ____ $=$ _____

The original square is at c3.

c3 $+$ $+$ $=$ ____ $+$ ____ $+$ ____ $=$ _____

Student Name _____ Date _____

Addition

e	11	12	13	14	15
d	16	17	18	19	20
c	21	22	8	23	24
b	25	26	27	28	29
a	30	31	32	33	34
	1	2	3	4	5

Move by one square at a time from the original square.

The original square is at c3.

C3 + = _____ + _____ = _____

The original square is at c3.

C3 + = _____ + _____ = _____

The original square is at c3.

C3 + = _____ + _____ = _____

The original square is at c3.

C3 + = _____ + _____ = _____

Student Name _____ Date _____

Addition

e	11	12	13	14	15
d	16	17	18	19	20
c	21	22	8	23	24
b	25	26	27	28	29
a	30	31	32	33	34
	1	2	3	4	5

Move by one square at a time from the original square.

The original square is at c3.

C3 ＋ ┼ ＝ _____ ＋ _____ ＝ _____

The original square is at c3.

C3 ＋ ┼ ＝ _____ ＋ _____ ＝ _____

The original square is at c3.

C3 ＋ ┼ ＝ _____ ＋ _____ ＝ _____

The original square is at c3.

C3 ＋ ┼ ＝ _____ ＋ _____ ＝ _____

Student Name _____ Date _____

Addition

e	11	12	13	14	15
d	16	17	18	19	20
c	21	22	9	23	24
b	25	26	27	28	29
a	30	31	32	33	34
	1	2	3	4	5

Move by one square at a time from the original square.

The original square is at c3.

C3 + = _____ + _____ = _____

The original square is at c3.

C3 + = _____ + _____ = _____

The original square is at c3.

C3 + = _____ + _____ = _____

The original square is at c3.

C3 + = _____ + _____ = _____

Addition

e	11	12	13	14	15
d	16	17	18	19	20
c	21	22	9	23	24
b	25	26	27	28	29
a	30	31	32	33	34
	1	2	3	4	5

Move by one square at a time from the original square.

The original square is at c3.

C3 + ⬍⬌ + ⬍⬌ = _____ + _____ + _____ = _____

The original square is at c3.

C3 + ⬍⬆ + ⬍⬆ = _____ + _____ + _____ = _____

The original square is at c3.

C3 + ⬍⬇ + ⬍⬇ = _____ + _____ + _____ = _____

The original square is at c3.

C3 + ⬍⬇ + ⬍⬇ = _____ + _____ + _____ = _____

Student Name _____ Date _____

Addition

e	11	12	13	14	15
d	16	17	18	19	20
c	21	22	9	23	24
b	25	26	27	28	29
a	30	31	32	33	34
	1	2	3	4	5

Move by one square at a time from the original square.

The original square is at c3.

C3 $+$ $+$ = _____ $+$ _____ $+$ _____ = _____

The original square is at c3.

C3 $+$ $+$ = _____ $+$ _____ $+$ _____ = _____

The original square is at c3.

C3 $+$ $+$ = _____ $+$ _____ $+$ _____ = _____

The original square is at c3.

C3 $+$ $+$ = _____ $+$ _____ $+$ _____ = _____

Student Name _____ Date _____

Addition

e	11	12	13	14	15
d	16	17	18	19	20
c	21	22	9	23	24
b	25	26	27	28	29
a	30	31	32	33	34
	1	2	3	4	5

Move by one square at a time from the original square.

The original square is at c3.

c3 $+$ ✕ = _____ $+$ _____ = _____

The original square is at c3.

c3 $+$ ✕ = _____ $+$ _____ = _____

The original square is at c3.

c3 $+$ ✕ = _____ $+$ _____ = _____

The original square is at c3.

c3 $+$ ✕ = _____ $+$ _____ = _____ 26

Addition

e	11	12	13	14	15
d	16	17	18	19	20
c	21	22	9	23	24
b	25	26	27	28	29
a	30	31	32	33	34
	1	2	3	4	5

Move by one square at a time from the original square.

The original square is at c3.

C3 $+ \times + \times =$ ____ $+$ ____ $+$ ____ $=$ _____

The original square is at c3.

C3 $+ \times + \times =$ ____ $+$ ____ $+$ ____ $=$ _____

The original square is at c3.

C3 $+ \times + \times =$ ____ $+$ ____ $+$ ____ $=$ _____

The original square is at c3.

C3 $+ \times + \times =$ ____ $+$ ____ $+$ ____ $=$ _____

Student Name _____ Date _____

Addition

e	11	12	13	14	15
d	16	17	18	19	20
c	21	22	9	23	24
b	25	26	27	28	29
a	30	31	32	33	34
	1	2	3	4	5

Move by one square at a time from the original square.

The original square is at c3.

c3 $+$ $+$ $=$ ___ $+$ ___ $+$ ___ $=$ ___

The original square is at c3.

c3 $+$ $+$ $=$ ___ $+$ ___ $+$ ___ $=$ ___

The original square is at c3.

c3 $+$ $+$ $=$ ___ $+$ ___ $+$ ___ $=$ ___

The original square is at c3.

c3 $+$ $+$ $=$ ___ $+$ ___ $+$ ___ $=$ ___

Addition

e	11	12	13	14	15
d	16	17	18	19	20
c	21	22	9	23	24
b	25	26	27	28	29
a	30	31	32	33	34
	1	2	3	4	5

Move by one square at a time from the original square.

The original square is at c3.

C3 + ┼ = _____ + _____ = _____

The original square is at c3.

C3 + ┼ = _____ + _____ = _____

The original square is at c3.

C3 + ┼ = _____ + _____ = _____

The original square is at c3.

C3 + ┼ = _____ + _____ = _____

Ho Math Chess 何数棋谜　妈！我会棋谜式加法啦！
Mom! I Learn Addition Using Math-Chess-Puzzles Connection
Contents include both traditional and Math-Chess-Puzzles combined methods. Extra strength

©2008 – 2018 Frank Ho, Amanda Ho　　All rights reserved. www.homathchess.com

Student Name _____ Date _____

Addition

e	11	12	13	14	15
d	16	17	18	19	20
c	21	22	9	23	24
b	25	26	27	28	29
a	30	31	32	33	34
	1	2	3	4	5

Move by one square at a time from the original square.

The original square is at c3.

C3 + = _____ + _____ = _____

The original square is at c3.

C3 + = _____ + _____ = _____

The original square is at c3.

C3 + = _____ + _____ = _____

The original square is at c3.

C3 + = _____ + _____ = _____

Addition and subtraction concept by link

3	6	3	7
2	2	11	9
1	4	8	5
	a	b	c

You are a chess piece and located at b2.

The original square is at the first term	The original square is at the first term
b2= ⬍ + ⬍ _____ = __+__	b2= ⬍ + ⬍ _____ = __+__
⬍ = b2 – ⬍ _____ = __–__	⬍ = b2 – ⬍ _____ = __–__
b2= ⬍ + ⬍ _____ = __+__	b2= ⬍ + ⬍ _____ = __+__
⬍ = b2 – ⬍ _____ = __–__	⬍ = b2 – ⬍ _____ = __–__

Addition and subtraction concept

3	6	5	7
2	8	14	6
1	7	9	8
	a	b	c

You are a chess piece and located at b2.

The original square is at the first term	The original square is at the first term
b2= + _____ = __+__ .	b2= + _____ = __+__
= b2 – _____ = __–__	= b2 – _____ = __–__
b2= + _____ = __+__	b2= + _____ = __+__
= b2 – _____ = __–__	= b2 – _____ = __–__

Addition and subtraction concept

3	9	6	5
2	8	13	5
1	8	7	4
	a	b	c

You are a chess piece and located at b2.

The original square is at the first term	The original square is at the first term
b2= ⊕ + ⊕ _____ = __+__	b2= ⊕ + ⊕ _____ = __+__
⊕ = b2 – ⊕ _____ = __–__	⊕ = b2 – ⊕ _____ = __–__
b2= ⊕ + ⊕ _____ = __+__	b2= ⊕ + ⊕ _____ = __+__
⊕ = b2 – ⊕ _____ = __–__	⊕ = b2 – ⊕ _____ = __–__

Student Name _____ Date _____

Addition and subtraction concept

3	9	7	6
2	8	12	4
1	6	5	3
	a	b	c

You are a chess piece and located at b2.

The original square is at the first term	The original square is at the first term
b2= ⬌⬍ + ⬌⬍ _____ = __+__	b2= ⬌⬍ + ⬌⬍ _____ = __+__
⬌⬍ = b2 − ⬌⬍ _____ = __−__	⬌⬍ = b2 − ⬌⬍ _____ = __−__
b2= ⬌⬍ + ⬌⬍ _____ = __+__	b2= ⬌⬍ + ⬌⬍ _____ = __+__
⬌⬍ = b2 − ⬌⬍ _____ = __−__	⬌⬍ = b2 − ⬌⬍ _____ = __−__

Addition and subtraction concept

3	6	6	9
2	7	15	8
1	6	9	9
	a	b	c

You are a chess piece and located at b2.

The original square is at the first term	The original square is at the first term
b2= ⬍ + ⬍ _____ = __+__	b2= ⬍ + ⬍ _____ = __+__
⬍ = b2 − ⬍ _____ = __−__	⬍ = b2 − ⬍ _____ = __−__
b2= ⬍ + ⬍ _____ = __+__	b2= ⬍ + ⬍ _____ = __+__
⬍ = b2 − ⬍ _____ = __−__	⬍ = b2 − ⬍ _____ = __−__

Addition and subtraction concept

3	9	9	8
2	8	17	9
1	9	8	8
	a	b	c

You are a chess piece and located at b2.

The original square is at the first term	The original square is at the first term
b2= ⬍ + ⬍ _____ = __+__	b2= ⬍ + ⬍ _____ = __+__
⬍ = b2 – ⬍ _____ = __–__	⬍ = b2 – ⬍ _____ = __–__
b2= ⬍ + ⬍ _____ = __+__	b2= ⬍ + ⬍ _____ = __+__
⬍ = b2 – ⬍ _____ = __–__	⬍ = b2 – ⬍ _____ = __–__

Addition and subtraction concept

3	9	7	8
2	9	16	7
1	8	9	7
	a	b	c

You are a chess piece and located at b2.

The original square is at the first term	The original square is at the first term
b2= + _____ = __+__	b2= + _____ = __+__
= b2 – _____ = __–__	= b2 – _____ = __–__
b2= + _____ = __+__	b2= + _____ = __+__
= b2 – _____ = __–__	= b2 – _____ = __–__

Student Name _____ Date _____

Addition and subtraction concept

3	9	9	9
2	9	18	9
1	9	9	9
	a	b	C

You are a chess piece and located at b2.

The original square is at the first term	The original square is at the first term
b2= + _____ = __+__	b2= + _____ = __+__
= b2 – _____ = __–__	= b2 – _____ = __–__
b2= + _____ = __+__	b2= + _____ = __+__
= b2 – _____ = __–__	= b2 – _____ = __–__

Addition and subtraction concept

3	6	3	7
2	2	11	9
1	4	8	5
	a	b	c

You are a chess piece and located at b2.

The original square is at the first term	The original square is at the first term
b2 = ⤬ + ⤬ _____ = __ + __	b2 = ⤬ + ⤬ _____ = __ + __
⤬ = b2 – ⤬ _____ = __ – __	⤬ = b2 – ⤬ _____ = __ – __
b2 = ⤬ + ⤬ _____ = __ + __	b2 = ⤬ + ⤬ _____ = __ + __
⤬ = b2 – ⤬ _____ = __ – __	⤬ = b2 – ⤬ _____ = __ – __

Student Name _____ Date _____

Addition and subtraction concept

3	6	5	7
2	8	14	6
1	7	9	8
	a	b	c

You are a chess piece and located at b2.

The original square is at the first term	The original square is at the first term
b2 = ⤢ + ⤢ _____ = __ + __	b2 = ⤢ + ⤢ _____ = __ + __
⤢ = b2 − ⤢ _____ = __ − __	⤢ = b2 − ⤢ _____ = __ − __
b2 = ⤢ + ⤢ _____ = __ + __	b2 = ⤢ + ⤢ _____ = __ + __
⤢ = b2 − ⤢ _____ = __ − __	⤢ = b2 − ⤢ _____ = __ − __

Addition and subtraction concept

3	7	6	5
2	4	13	9
1	8	7	6
	a	b	c

You are a chess piece and located at b2.

The original square is at the first term	The original square is at the first term
b2 = ✕ + ✕ _____ = __ + __	b2 = ✕ + ✕ _____ = __ + __
✕ = b2 – ✕ _____ = __ – __	✕ = b2 – ✕ _____ = __ – __
b2 = ✕ + ✕ _____ = __ + __	b2 = ✕ + ✕ _____ = __ + __
✕ = b2 – ✕ _____ = __ – __	✕ = b2 – ✕ _____ = __ – __

Student Name _____ Date _____

Addition and subtraction concept

3	9	7	6
2	8	12	4
1	6	5	3
	a	b	c

You are a chess piece and located at b2.

The original square is at the first term	The original square is at the first term
b2 = ✗ + ✗ _____ = __ + __	b2 = ✗ + ✗ _____ = __ + __
✗ = b2 – ✗ _____ = __ – __	✗ = b2 – ✗ _____ = __ – __
b2 = ✗ + ✗ _____ = __ + __	b2 = ✗ + ✗ _____ = __ + __
✗ = b2 – ✗ _____ = __ – __	✗ = b2 – ✗ _____ = __ – __

Student Name _____ Date _____

Addition and subtraction concept

3	6	6	9
2	7	15	8
1	6	9	9
	a	b	c

You are a chess piece and located at b2.

The original square is at the first term	The original square is at the first term
b2 = ⤬ + ⤬ _____ = __ + __	b2 = ⤬ + ⤬ _____ = __ + __
⤬ = b2 – ⤬ _____ = __ – __	⤬ = b2 – ⤬ _____ = __ – __
b2 = ⤬ + ⤬ _____ = __ + __	b2 = ⤬ + ⤬ _____ = __ + __
⤬ = b2 – ⤬ _____ = __ – __	⤬ = b2 – ⤬ _____ = __ – __

Addition and subtraction concept

3	7	8	9
2	8	16	8
1	7	9	9
	a	b	c

You are a chess piece and located at b2.

The original square is at the first term	The original square is at the first term
b2 = + _____ = __ + __	b2 = + _____ = __ + __
= b2 − _____ = __ − __	= b2 − _____ = __ − __
b2 = + _____ = __ + __	b2 = + _____ = __ + __
= b2 − _____ = __ − __	= b2 − _____ = __ − __

Addition and subtraction concept

3	8	8	9
2	9	17	8
1	8	9	9
	a	b	c

You are a chess piece and located at b2.

The original square is at the first term	The original square is at the first term
b2 = ⤬ + ⤬ _____ = __ + __	b2 = ⤬ + ⤬ _____ = __ + __
⤬ = b2 − ⤬ _____ = __ − __	⤬ = b2 − ⤬ _____ = __ − __
b2 = ⤬ + ⤬ _____ = __ + __	b2 = ⤬ + ⤬ _____ = __ + __
⤬ = b2 − ⤬ _____ = __ − __	⤬ = b2 − ⤬ _____ = __ − __

Addition and subtraction concept

3	9	9	9
2	9	18	9
9	6	9	9
	a	b	c

You are a chess piece and located at b2.

The original square is at the first term	The original square is at the first term
b2 = ⤬ + ⤬ _____ = __ + __	b2 = ⤬ + ⤬ _____ = __ + __
⤬ = b2 – ⤬ _____ = __ – __	⤬ = b2 – ⤬ _____ = __ – __
b2 = ⤬ + ⤬ _____ = __ + __	b2 = ⤬ + ⤬ _____ = __ + __
⤬ = b2 – ⤬ _____ = __ – __	⤬ = b2 – ⤬ _____ = __ – __

Addition and subtraction by link using dots

5		•• -1		•• ••	
4	•• +1	•• ••	••	••	••
3		••• +2		•• +3	
2	• +2	•• ••	•• -1	•• ••	••
1		•		• •• +3	
	a	b	c	d	e

Inside row 3, column c box:

2+1	5+1	6+2
7+3	3+2	9+2
8+1	5+4	4+3

You are a chess piece located at c3

• = 1

Addition and subtraction by link

5		•• ••		•• ••	
4	•• •	2+ •• •••	•• +3	••	•• •• +4
3		• • •	5-3 6+2 3-1 / 8+3 3+2 7-4 / 5+2 7+2 6-1	•	
2	• +3	0+ •• •••	•• •• +5	•• ••	•• ••
1		•		3+ •• •	
	a	b	c	d	e

You are a chess piece located at c3 .
• = 1

Student Name _____ Date _____

Addition and subtraction by link

5		2+				
4			+1			
3		4+	9+3 8-2 3+2 / 7+4 4-3 6+3 / 4+3 6-4 5-2			
2		2+	+3			
1		+3				
	a	b	c		d	e

You are a chess piece located at c3.

● = 1

Ho Math Chess 何数棋谜　妈！我会棋谜式加法啦！
Mom! I Learn Addition Using Math-Chess-Puzzles Connection
Contents include both traditional and Math-Chess-Puzzles combined methods. Extra strength

©2008 – 2018 Frank Ho, Amanda Ho　　All rights reserved. www.homathchess.com

Student Name _____ Date _____

Addition and subtraction by link

5		●●		●●●	
4	●●● +4	●●● ●●	●● -1	●●	2+ ●●
3		●●●	1+2 4+7 4+6 / 3+7 2+3 3+4 / 2+6 2+5 1+5	●	
2	● +5	●●● ●●	●●● -1	●● ●●	●● +3
1		●		●●●	
	a	b	c	d	e

You are a chess piece located at c3 ⊞ .

● = 1

Student Name _____ Date _____

Addition and subtraction by link

5		●● ●●		●● ● ●●	
4	●● ●	●● ● ●● +1	●● +5	●●	3+ ●● ●●
3		● ● ●	4-2 1+5 6-3 / 2+6 5+5 6+3 / 3+6 8-2 5-5	●	
2	●	●● ●● +3	●● ● +5	●● ●● +3	●● ●
1		●		●● ● +2	
	a	b	c	d	e

You are a chess piece located at c3 ⊞.
● = 1

Mom! I Learn Addition Using Math-Chess-Puzzles Connection

Contents include both traditional and Math-Chess-Puzzles combined methods. Extra strength

Student Name _____ Date _____

Addition and subtraction by link

5		•• ••		••• •• +3	
4	5+ ••	•• •• •• +3-3	•• +9	••	•• ••
3		•••	3+3 7-5 6-4 / 4+4 9-9 5+3 / 7-5 8-7 9+0	•	
2	6+ •	•• •• - 4	•• •• -2	•• •• +3	•••
1		• +3		•• •• +2	
	a	b	c	d	e

You are a chess piece located at c3 [ ].

• = 1

Addition and subtraction by link

5		••			••	
4	••	•••		••	••	
3		•••	7+3 6-5 4-3 / 7+4 5-4 3-1 / 6-4 9-2 2+8		•	
2	•	•••	••	••	•••	
1		•		••		
	a	b	c	d	e

You are a chess piece located at c3 □.
• = 1

Addition and subtraction by link

5		•• ••			•• ••	
4	•• •	•• •• +4	•• +5		••	•• •• +4
3		•••	3+6 2+7 9-3 / 1+5 8-6 6+4 / 7-5 5+4 4+5		•	
2	•	•• •• +5	•• •• +5		•• ••	•• • +5
1		•			•• •	
	a	b	c		d	e

You are a chess piece located at c3

• = 1

d + d of math and chess integrated problems

One-digit 5 addition

5		3		4	
4	5	3	2	4	5
3		2	5	3	
2	3	4	3	5	4
1		5		2	
	a	b	c	d	e

You are at square c3 = ☐.

6 addition

5		4		3	
4	5	5	2	4	5
3		4	6	5	
2	5	6	3	6	2
1		4		3	
	a	b	c	d	e

You are at square c3 = ☐ .

+ = ____	+ = ____
+ = ____	+ = ____
+ = ____	+ = ____
+ = ____	+ = ____
+ = ____	+ = ____
+ = ____	+ = ____
+ = ____ 10	+ = ____ 11
+ = ____ 11	+ = ____ 10

7 addition

5		4		3	
4	5	5	2	4	5
3		4	7	5	
2	5	6	3	6	2
1		4		3	
	a	b	c	d	e

You are at square c3 = ☐ .

Student Name _____   Date _____

8 addition

5		4		3	
4	5	5	2	4	5
3		4	8	5	
2	5	6	3	6	2
1		4		3	
	a	b	c	d	e

You are at square c3 = ☐.

Student Name _____ Date _____

9 addition

5		3		4	
4	5	6	3	2	5
3		2	9	3	
2	3	5	4	4	4
1		5		2	
	a	b	c	d	e

You are at square c3 = ☐.

11 addition

5		13		14	
4	15	16	13	12	15
3		19	11	13	
2	13	18	14	4	14
1		15		19	
	a	b	c	d	e

You are at square c3 = ☐.

12 addition

5		13		14	
4	15	14	12	13	15
3		19	12	15	
2	13	18	16	4	14
1		17		19	
	a	b	c	d	e

You are at square c3 = ☐.

13 addition

5		13		14	
4	15	16	13	12	15
3		19	13	13	
2	13	18	14	4	14
1		15		19	
	a	b	c	d	e

You are at square c3 = ☐.

Adding (be patient with your writing.)

3	3090807	4060706	4070409
2	3040506	3060504	3040708
1	4090709	2070608	3090709
	a	b	c

You are at square b2

= ☐ .

Adding (be patient with your writing.)

3	3090807	4060706	4070409
2	3040506	3060504	3040708
1	4090709	2070608	3090709
	a	b	c

You are at square b2

= ☐ .

Ho Math Chess 何数棋谜　妈！我会棋谜式加法啦！

Mom! I Learn Addition Using Math-Chess-Puzzles Connection

Contents include both traditional and Math-Chess-Puzzles combined methods. Extra strength

©2008 － 2018 Frank Ho, Amanda Ho　　All rights reserved. www.homathchess.com

Student Name _____ Date _____

Adding (be patient with your writing.)

3	3090807	4060706	4070409
2	3040506	3060504	3040708
1	4090709	2070608	3090709
	a	b	c

You are at square b2

= ☐ .

Part 5: Cognitive ability 思唯意識

Puzzle # 1

Complete the sequence.

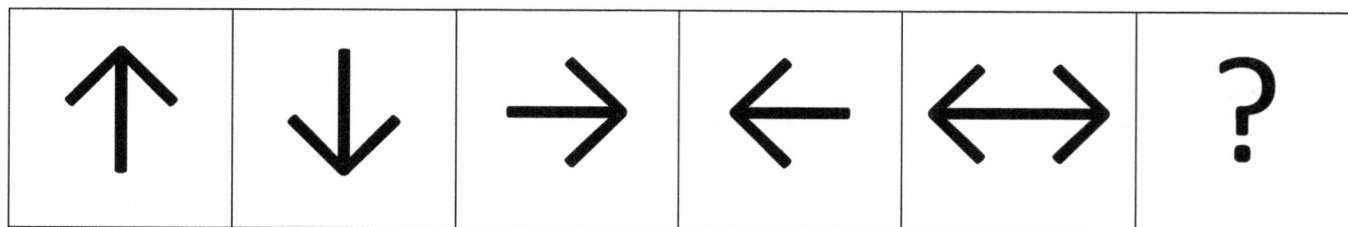

↑	↓	→	←	↔	?

Puzzle # 2

Complete the sequence.

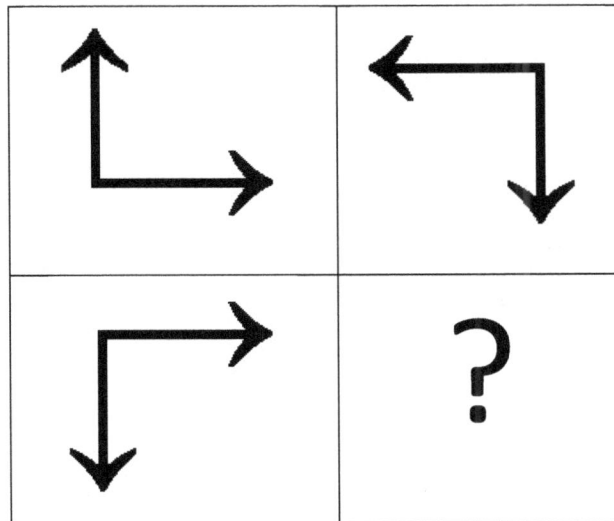

Student Name _____ Date _____

Puzzle # 3

□ = **5**

△ = **3**

◯ − ⊡ = **5**

◯ is a whole number from 0 to 9.

What is ◯ ?

Puzzle # 4

Replace each ? with a number.

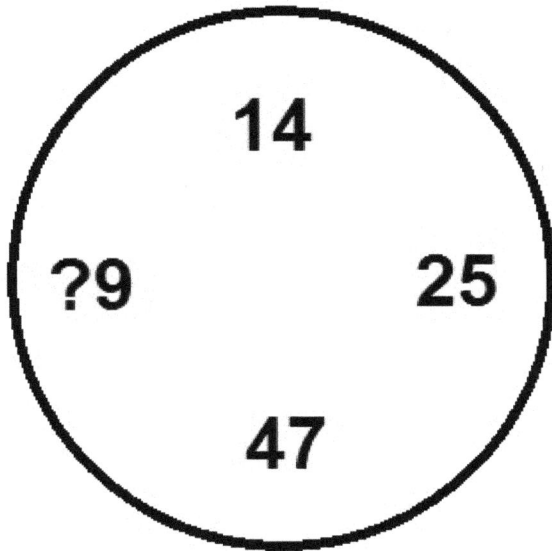

Puzzle # 5

Replace ? by a number.

1, 1, 2, 3, ?, 8, 13

Student Name _____ Date _____

Puzzle # 6

Replace ? by a number.

1.

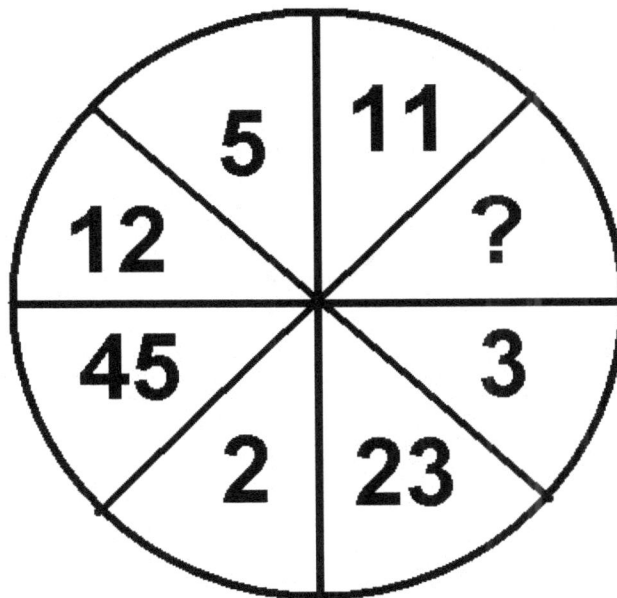

Puzzle # 7

Replace ? by a number.

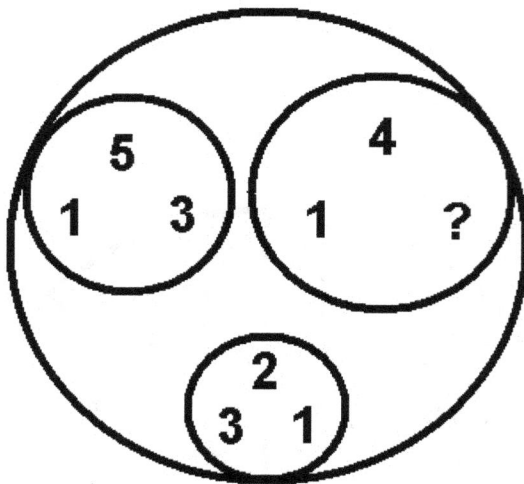

Puzzle # 8

Replace each ? with a number.

1, 22, 333, 444?, 55?55, ?

Puzzle # 9

Replace each ? with a number.

0	0
0	1
1	0
?	?

Replace each ? with a letter.

H	T
T	H
T	T
?	?

Puzzle # 10

Replace ? by a number.

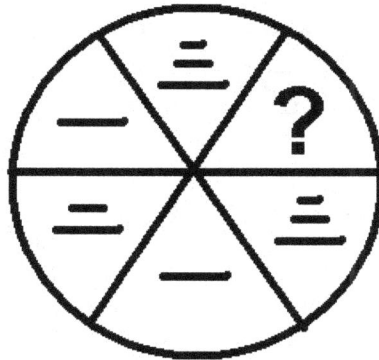

Puzzle # 11

What is the next term in the following sequence?

3 + 1, 6 + 3 + 1, 10 + 6 + 3 + 1, _____
3.

One-drawable figure

Circle the following figure which can be traced with one continuous line without lifting your pencil or retracing it.

Find the answer for the ?

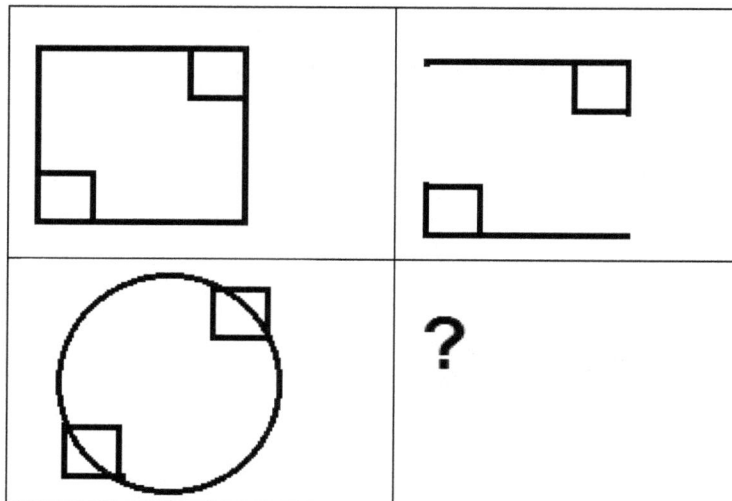

Student Name _____ Date _____

Find the answer for the ?

□	⊞
△	?

Matrix Reasoning

Number or figure pattern Completion

Draw the missing piece.

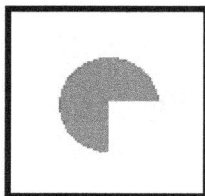

Answer _____

Reasoning by Analogy (row or column change)

Find the answer for ? by examining how the objects change across the row, column, or diagonal.

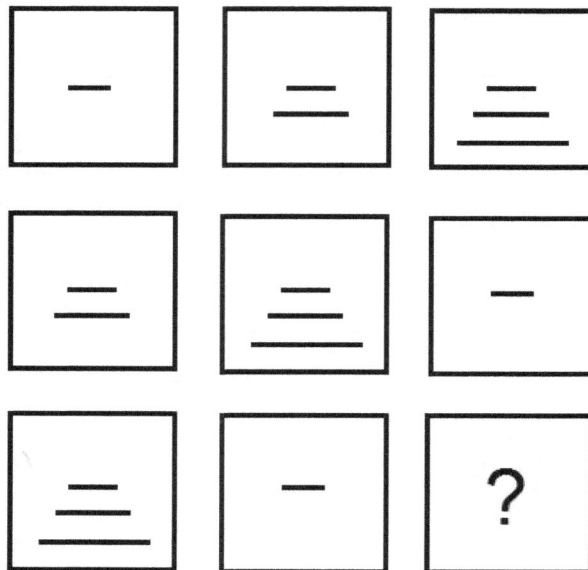

Answer _____

Matrix Reasoning

Serial Reasoning (row, column, and diagonal changes)

Find the answer for ? by examining how the objects change across the row and/or column and/or diagonal.

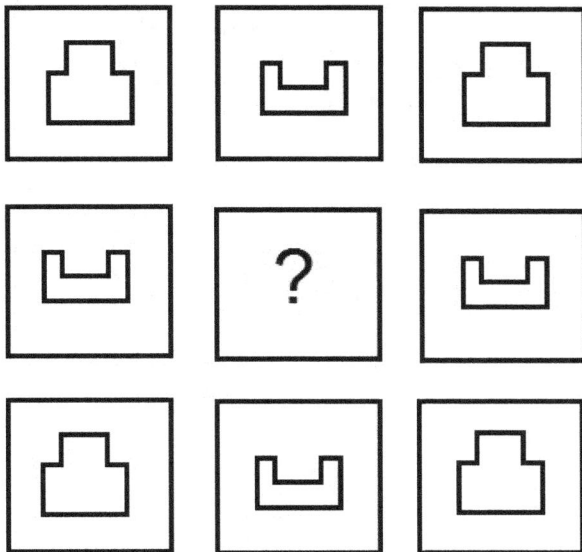

Answer _____

Spatial Reasoning (partial figures or numbers to make a whole or vice versa. Shape transformation may be involved.)

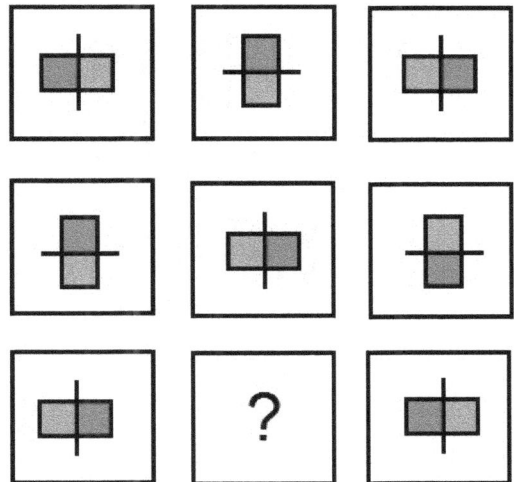

Answer _____

Matrix Reasoning

Number or figure pattern Completion	Reasoning by Analogy (row or column change)
Draw the missing piece.	Find the answer for ? by examining how the objects change across the row, column, or diagonal.

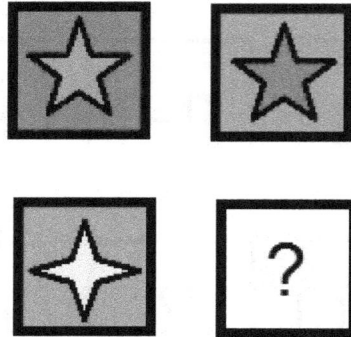

Answer _____

Answer _____

Student Name _____　Date _____

Matrix Reasoning

Serial Reasoning (row, column, and diagonal changes)

Find the answer for ? by examining how the objects change across the row and/or column and/or diagonal.

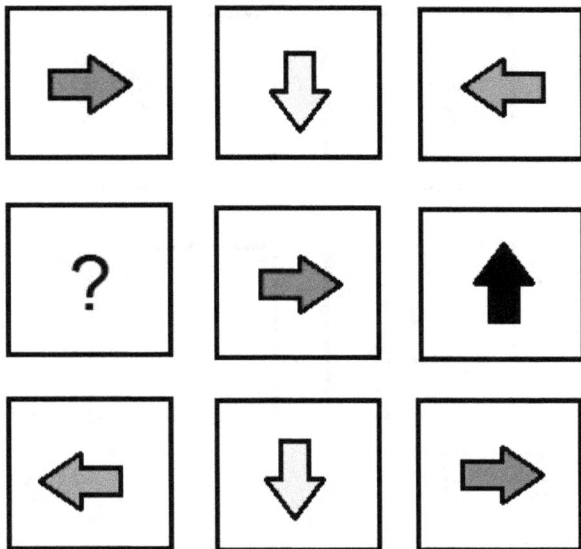

Answer _____

Spatial Reasoning (partial figures or numbers to make a whole or vice versa. Shape transformation may be involved.)

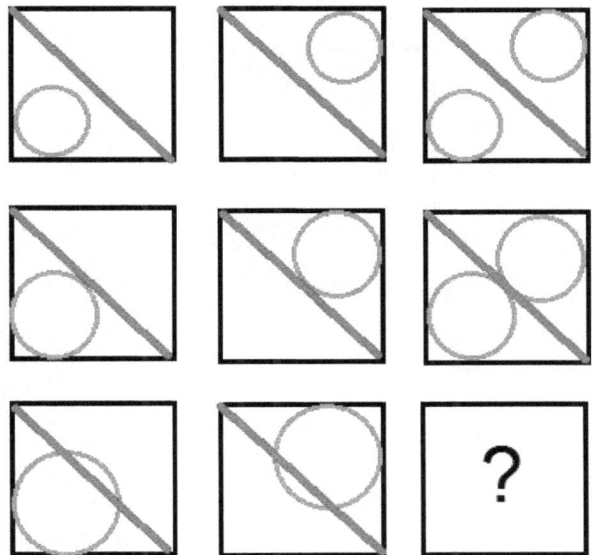

Answer _____

Student Name _____ Date _____

Matrix Reasoning

Number or figure pattern Completion	Reasoning by Analogy (row or column change)
Draw the missing piece.	Find the answer for ? by examining how the objects change across the row, column, or diagonal.
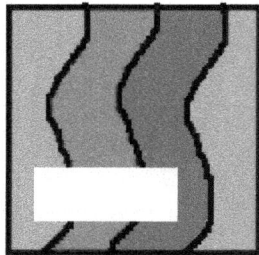	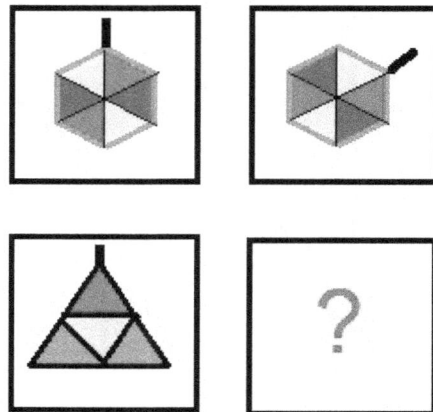
Answer _____	Answer _____

Student Name _____　Date _____

Matrix Reasoning

Serial Reasoning (row, column, and diagonal changes)

Find the answer for ? by examining how the objects change across the row and/or column and/or diagonal.

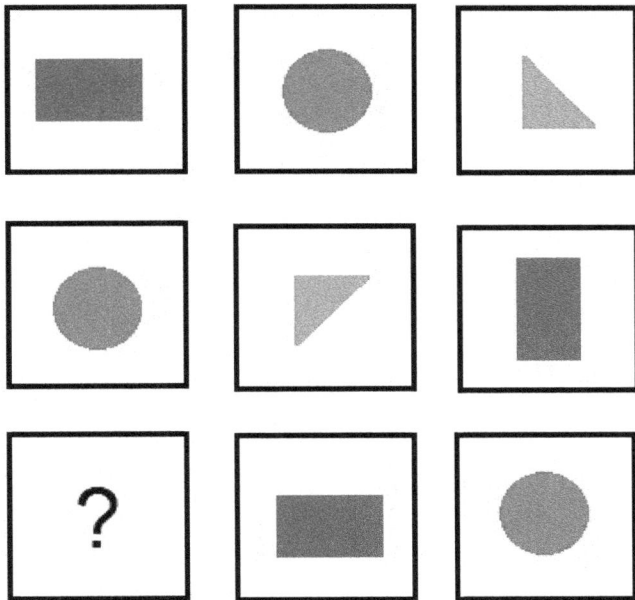

Answer _____

Spatial Reasoning (partial figures or numbers to make a whole or vice versa. Shape transformation may be involved.)

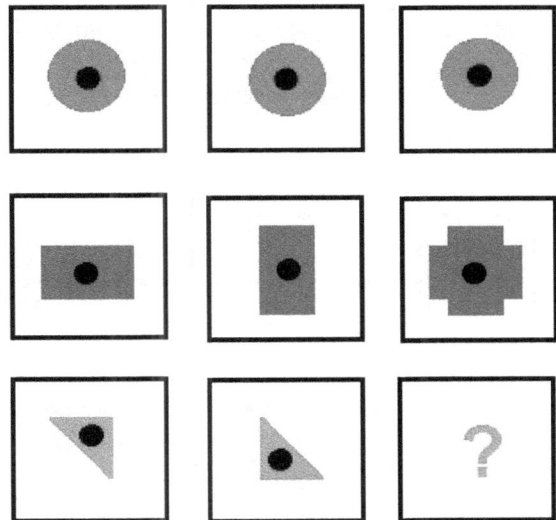

Answer _____

Matrix reasoning

Find an answer to each question mark.

Student Name ______________________________ Date ____________________

Matrix reasoning

Find an answer to each question mark.

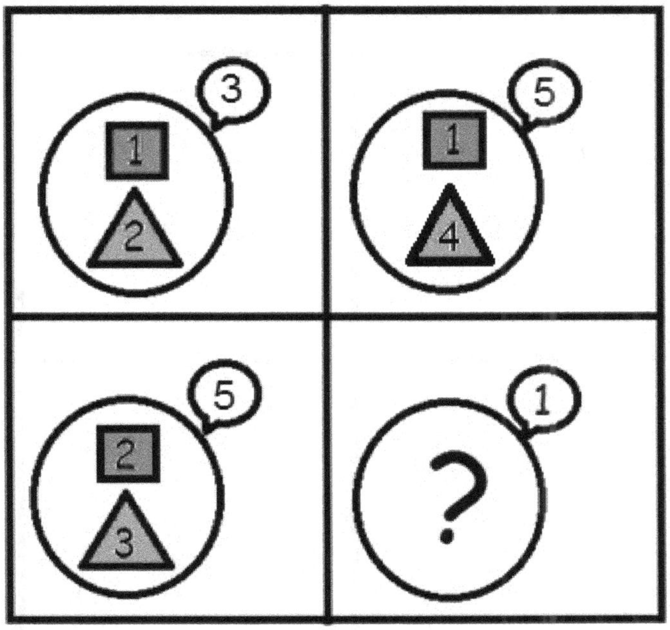

Matrix reasoning

Find an answer to each question mark.

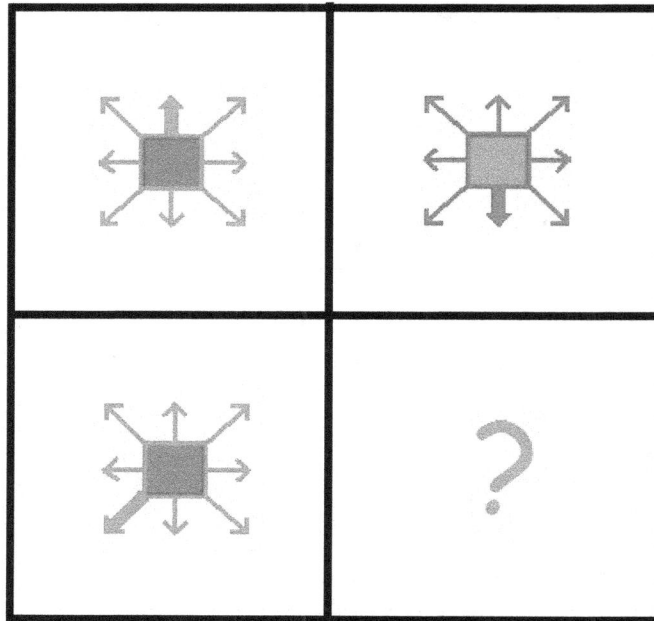

Matrix reasoning

Find an answer to each question mark.

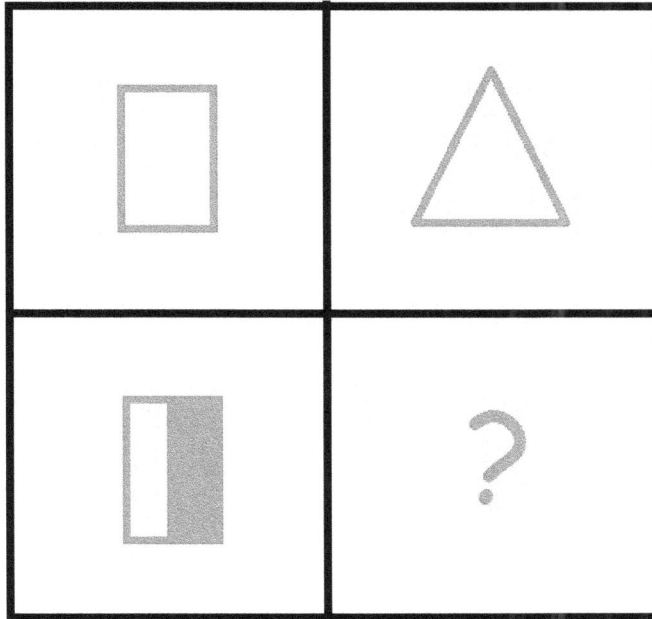

Student Name _____ Date _____

Matrix reasoning

Find an answer to each question mark.

Student Name ______________________________ Date ____________________

Matrix reasoning

Find an answer to each question mark.

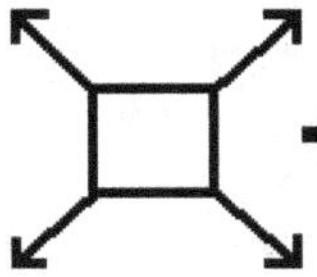 + 1	3 + 1
?	3 − 1

Student Name ______________________________ Date ____________________

Matrix reasoning

Find an answer to each question mark.

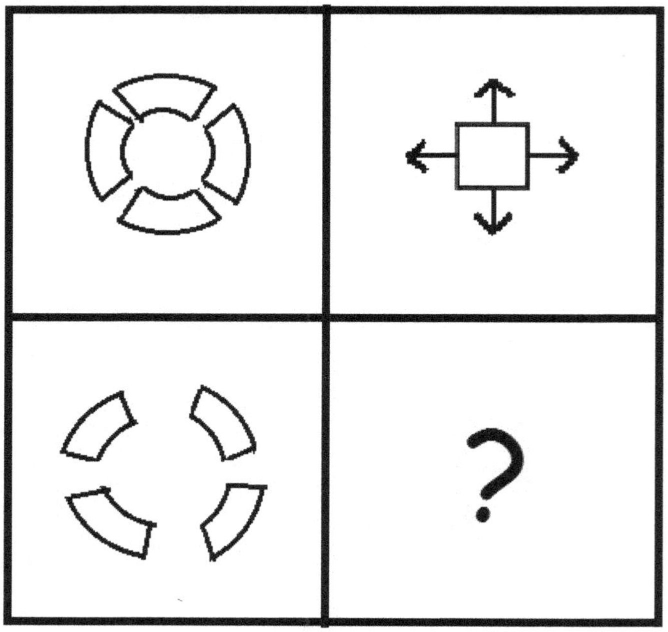

Part 6: Pattern, geometry, time, calendar, a fraction

Find the pattern and fill in each blank.

1	2	3	_____	_____	_____
2	4	6	_____	_____	_____
5	10	15	_____	_____	_____
10	100	1000	_____	_____	_____
325	335	345	_____	_____	_____
4236	5236	6236	_____	_____	_____
270	280	290	_____	_____	_____
1150	1200	1250	_____	_____	_____
7385	7375	7365	_____	_____	_____
4350	4250	4150	_____	_____	_____
1007	1008	1009	_____	_____	_____
4256	4506	4756	_____	_____	_____
21	32	43	_____	_____	_____
613	524	435	_____	_____	_____
87	76	65	_____	_____	_____

Looking for patterns

3	_____	15	31	63	127	255
10	20	40	80	160	_____	
2	4	16	_____	65536		
16	8	_____	_____	1		
_____	12	36	108	324	972	
_____	_____	20	24	96	100	400
600	600	300	100	25	_____	
784529	78452	7452	452	_____	_____	
11	18	25	32	39	_____	53
810	270	_____	30	10		
1	4	16	64	256	_____	
$400	$200	$100	$50	_____		
100	99	97	94	90	_____	_____
15	12	14	11	13	_____	_____
1	3	7	15	31	_____	_____
67	35	19	11	7	_____	_____

Looking for patterns.

```
                        1
                      2   2
                    3   4   3
                  4   7  __  4
                5  11  __  11   5
              6  __  25  __  16   6
            7  22  41  __  41  __  7
```

```
                        1
                      2   4
                    3   9   27
                  4  16  __  256
                5  __  __  625  3125
```

```
                        1
                      2   4
                    3   6   9
                  4   8  __  16
                5  __  15  __  __
              6  __  __  __  __  __
```

Complete the following Pattern.

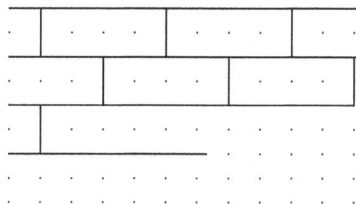

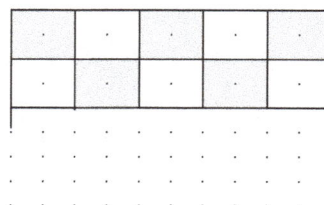

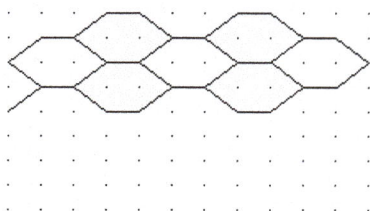

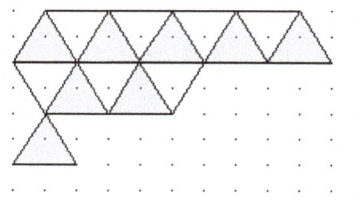

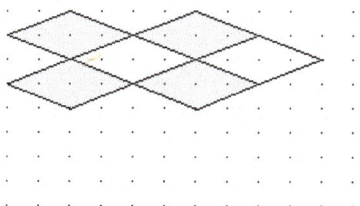

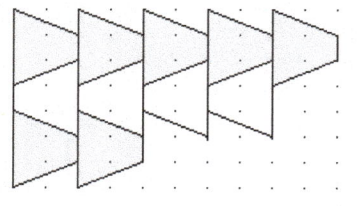

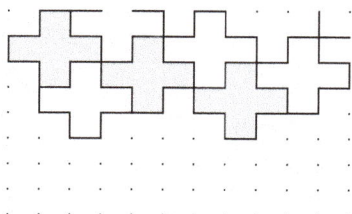

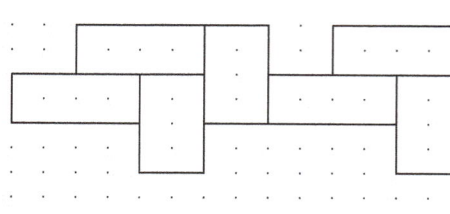

Part 7: Geometry

Lines

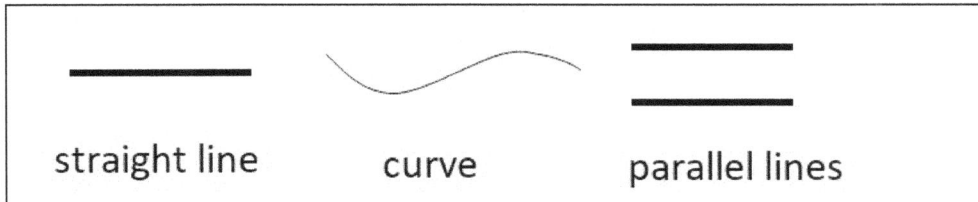

Matching.

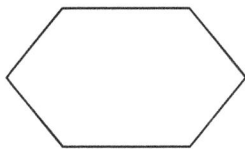   Straight lines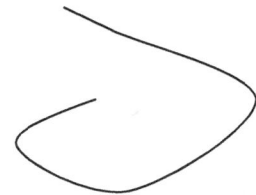

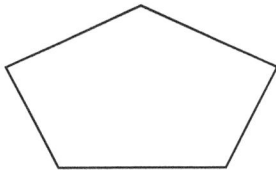   Parallel lines

   Curve

What kind of lines does each picture have?

	Straight line	Curve	Parallel lines

Draw line(s) through the following dots.

Straight line	Parallel lines	Curve
· 　B · A	· C 　· D	·　　· E　　F

Name of lines

horizontal lines	vertical lines	parallel lines
intersecting lines	perpendicular lines	

What types of lines can you find in each figure?

	horizontal lines	vertical lines	parallel lines	intersecting lines	perpendicular lines

Angles

An angle is formed by two rays with a common endpoint. An angle with a measure of 90 degrees is a right angle.

1. Arrange the angles in order of size from the largest to the smallest. Put in 1, 2, and 3 to show.

a.

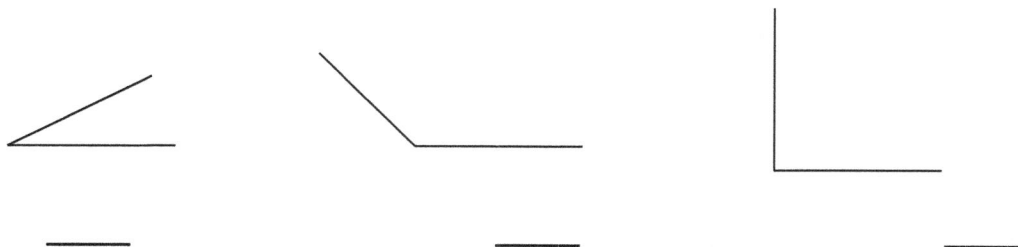

_____　　　　　_____　　　　　_____

b.

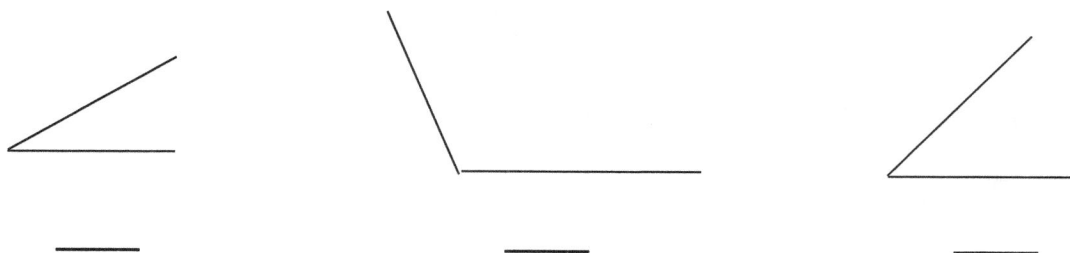

_____　　　　　_____　　　　　_____

2. Find the number of right angles of the following figures.

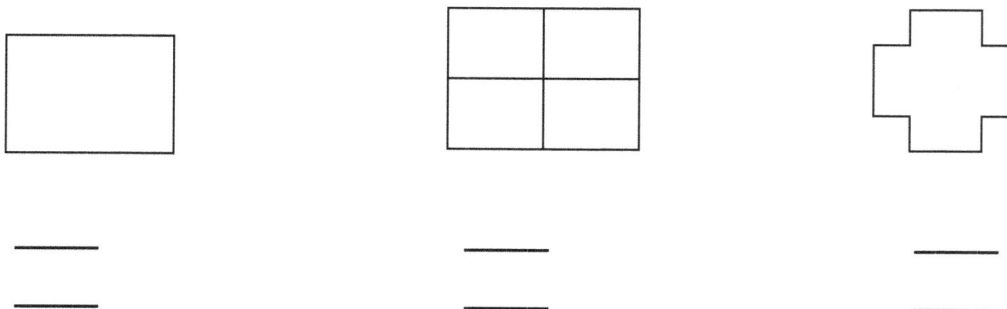

_____　　　　　_____　　　　　_____

_____　　　　　_____　　　　　_____

Student Name _____ Date _____

Shapes

Circle	Triangle (Three sides)	Rectangle (Four sides)
Pentagon (Five sides)	Hexagon (Six sides)	Octagon (Eight sides)

What types of shapes can you find in each figure?

	Circle	Triangle	Rectangle	Pentagon	Hexagon	Octagon

Student Name _____ Date _____

Quadrilaterals

Square	Rectangle	Rhombus
Parallelogram	Trapezoid	Kite

1.   How many rectangles can you find?

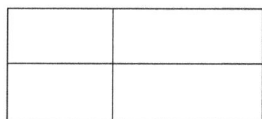

2.   How many triangles can you find?

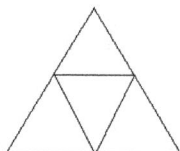

3.   How many squares can you find?

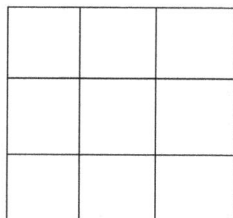

4.　　Matching.

Kite

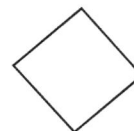

Square

Trapezoid

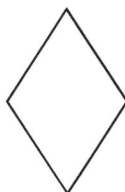

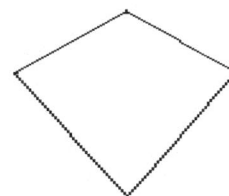

Rectangle

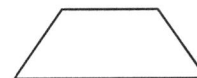

Rhombus

Parallelogram

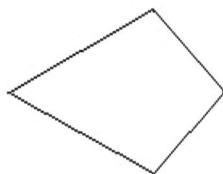

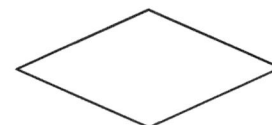

Shapes

The figure below is Chinese Tangram. It is made up seven pieces. Observe carefully and fill in the blanks below.

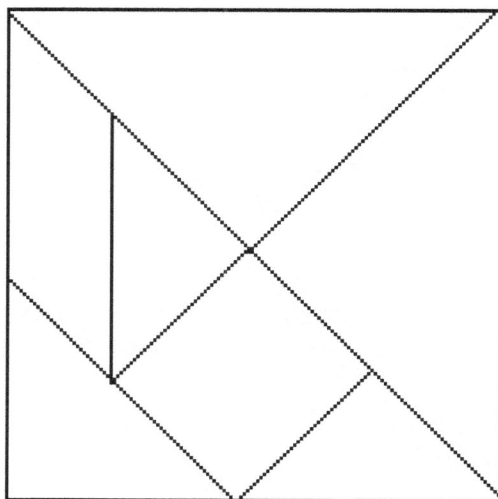

There are _____ triangles in the graph.

There are _____ squares in the graph.

There are _____ parallelograms in the graph.

There are _____ trapezoids in the graph.

Symmetry line

Find the lines of symmetry for these drawings

Part 8: Time and clock

Time

Telling time

1.

____o'clock

____ : ____

2.

____o'clock

____ : ____

3.

____o'clock

____ : ____

4.

____o'clock

____ : ____

5.

____o'clock

____ : ____

6.

____o'clock

____ : ____

7.

____o'clock

____ : ____

8.

____o'clock

____ : ____

9.

____o'clock

____ : ____

Student Name _____ Date _____

Telling time

1.

half past _____

_____ : _____

2.

half past _____

_____ : _____

3.

half past _____

_____ : _____

4.

Quarter past _____

_____ : _____

5.

Quarter past _____

_____ : _____

6.

Quarter past _____

_____ : _____

7.

Quarter to _____

_____ : _____

8.

Quarter to _____

_____ : _____

9.

Quarter to _____

_____ : _____

Telling time

1.

____ minutes (past/to) ____

____ : ____

2.

____ minutes (past/to) ____

____ : ____

3.

____ minutes (past/to) ____

____ : ____

4.

____ minutes (past/to) ____

____ : ____

5.

____ minutes (past/to) ____

____ : ____

6.

____ minutes (past/to) ____

____ : ____

7.

____ minutes (past/to) ____

____ : ____

8.

____ minutes (past/to) ____

____ : ____

9.

____ minutes (past/to) ____

____ : ____

Drawing the hands on the clock to show the time

1.

2:00

2.

4:00

3.

5:00

4.

8:00

5.

12:00

6.

3:00

7.

6:00

8.

9:00

9.

10:00

Student Name _____ Date _____

Drawing the hands on the clock to show the time.

1.

2:30

2.

half past 5

3.

9:30

4.

quarter past 5

5.

3:15

6.

10:15

7.

1:45

8.

quarter to 4

9.

quarter to 12

Drawing the hands on the clock to show the time.

1.

2:05

2.

7:15

3.

12:10

4.

9:35

5.

1:40

6.

6:20

7.

5 minutes past 7

8.

10 minutes to 8

9.

10 minutes after 5

Digital Time

| 01 : 25 am |
___ minutes past/to ___
in the morning/afternoon

| 09 : 55 am |
___ minutes past/to ___
in the morning/afternoon

| 03 : 40 pm |
___ minutes past/to ___
in the morning/afternoon

| 10 : 50 pm |
___ minutes past/to ___
in the morning/afternoon

| 11 : 45 am |
___ minutes past/to ___
in the morning/afternoon

| 04 : 25 pm |
___ minutes past/to ___
in the morning/afternoon

| 07 : 55 pm |
___ minutes past/to ___
in the morning/afternoon

| 06 : 10 am |
___ minutes past/to ___
in the morning/afternoon

| 08 : 40 am |
___ minutes past/to ___
in the morning/afternoon

Calendar

Fill in the following blank.

There are _____ months in a year.

In a regular year, February has _____ days. A regular year has _____ days.

In a leap year, February has _____ days. A leap year has _____ days.

There are about _____ weeks in a year.

January, March, May, July, August, October and December have _____ days each.

April, June, September and November have _____ days each.

There are _____ days in a week. They are Monday, _____, _____,

_____, _____, _____ and _____. I like _____ the best.

Today is January 6th. Tomorrow will be _____. The day after tomorrow

will be _____.

9. Today is March 10th. Yesterday was _____. The day before yesterday

was _____.

Part 9: Fraction

Example: Divide 1 whole into equal parts.

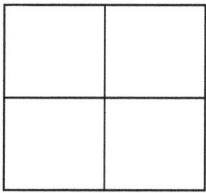

4 equal parts
4 fourth or 4 quarters

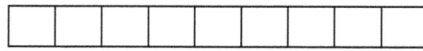

9 equal parts
5 fifth

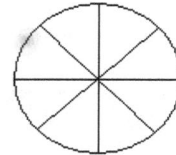

8 equal parts
10 tenth

1. Circle the figures that show equal parts.

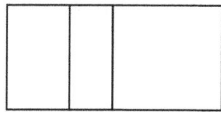

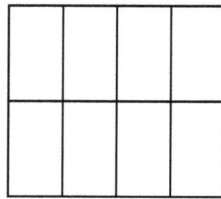

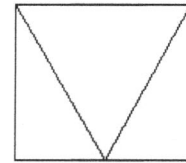

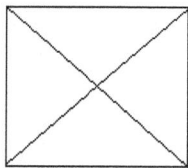

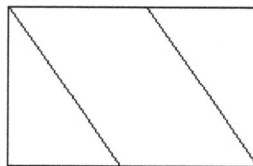

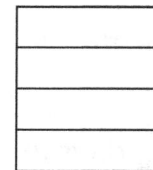

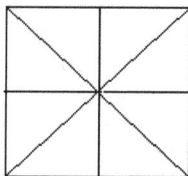

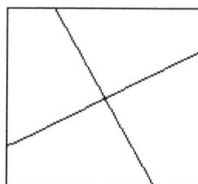

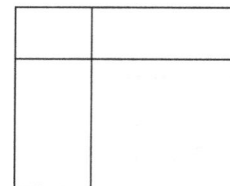

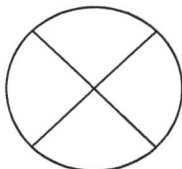

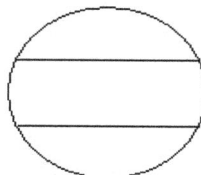

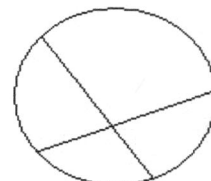

2. Name the equal parts of each whole.

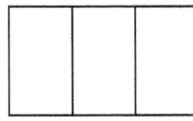

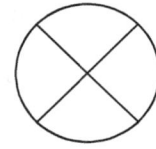

2 halves

_____

_____

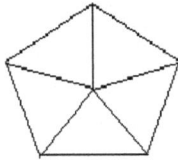

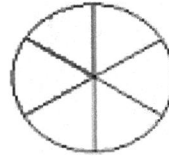

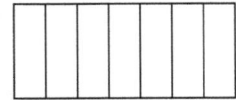

_____

_____

_____

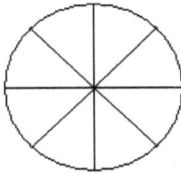

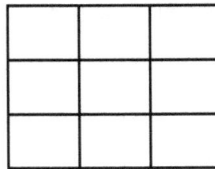

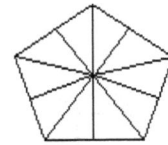

_____

_____

_____

).

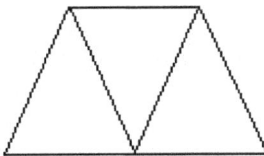

ι.

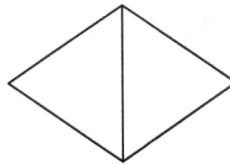

ι.

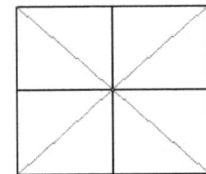

_____

_____

_____

Student Name _____ Date _____

Equal Parts

1. Name the equal parts of each whole.

Example:

　　3 equal parts　　　　　3 thirds

1. 　　2.

3. 　　4.

5. 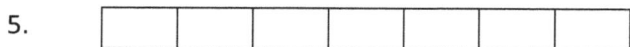　　6.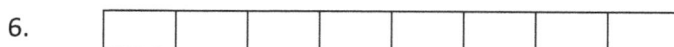

2. Divide each square into 2 halves in five ways.

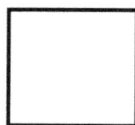 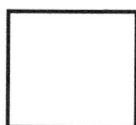 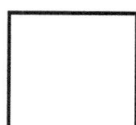  

3. Divide each square into four fourths in five ways.

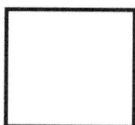 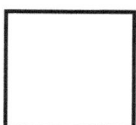 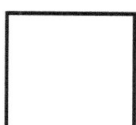 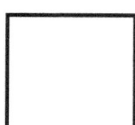 

Representing Fractions

$$\frac{\text{numerator}}{\text{denominator}} = \frac{\text{the number of favorite parts}}{\text{the number of equal parts in a whole}}$$

1. What fraction of each figure is shaded?

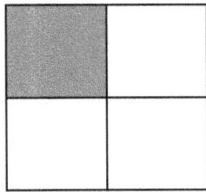

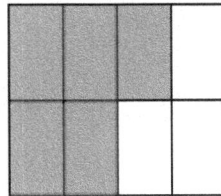

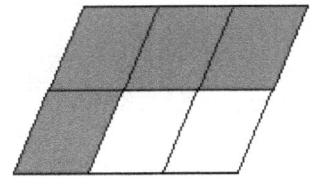

_____

_____

_____

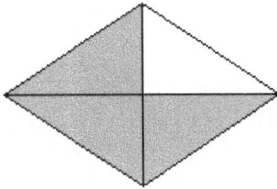

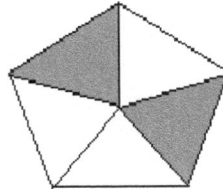

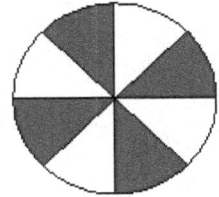

_____

_____

_____

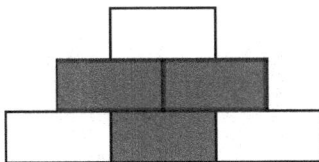

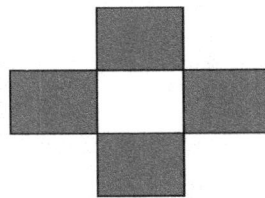

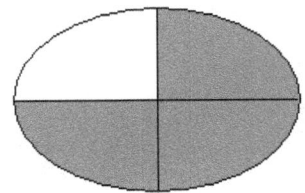

_____

_____

_____

2. What fraction of each figure is shaded? Express in words and in fraction form.

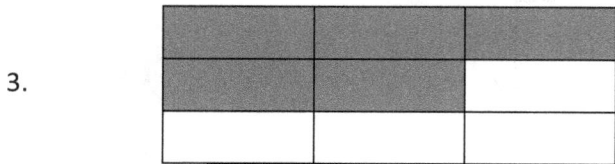

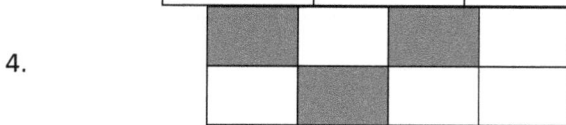

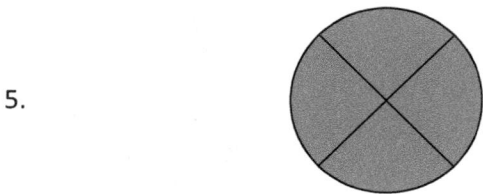

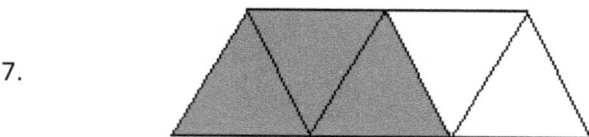

1.　　　　　　　　　　　　　　two thirds　　　　$\dfrac{2}{3}$

2.

3.

4.

5.

6.

7.

Part 10: Word problem solving

Each hand has five fingers. I count 20 fingers. How many hands are there?

Each car can carry 5 people. There are 3 cars. How many people can the car take?

One marble is 5¢. Ashley bought 4 marbles. How much did she spend on the marbles?

Write from 1 to 40. How many times do you write digit 5?

One nickel is 5¢. Linda has 20¢ in nickels. How many nickels does she have?

Ben has 25 chocolates. He divides them into a group of 5. How many groups can he make?

A rabbit eats 5 carrots each day. There are 30 carrots. Is it enough to feed the rabbit for 6 days?

Part 11: Tests

Test 1

If ◯ = 3 then ◯ + 1 = _____ .

If ◯ = 4 then ◯ + 1 = _____ .

If ⬡ = 1 then ◯ = ⬡⬡⬡ then ◯ + ⬡ = _____ .

If ◯ = ⬡⬡⬡ then ◯ + ⬡⬡ = _____ 5

If ◯ = ⬡⬡⬡ and ⬡ = 1 then ◯ – ⬡ = _____

Match left to right by drawing lines.

⬡	4
⬡⬡⬡⬡	2
⬡⬡	3
⬡⬡⬡	1

Out of 1, 2, and 3, what numbers added together will give answers of 3?

A number is drawn as follows to show 1 + 1 = 2.

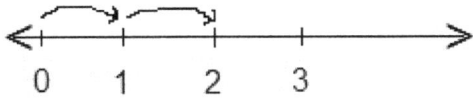

Look at the following diagram and come up with your equation.

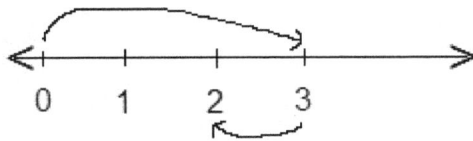

If the following diagram shows 3 - 1 = 2,

Then what is the value of the following diagram?

What is the value of the following diagram?

$37 - 9 =$

$73 - 4 =$

$45 - 8 =$

$89 + 7 =$

$36 - 17 =$

What is the largest one-digit number?

What is the largest two-digit number?

What is the smallest two-digit number?

Circle the following even numbers.

177, 46, 8, 91, 32, 21

Insert >, <, or = in ☐.

$(67 - 19)$ ☐ 47	$(31 - 18)$ ☐ 12
$(77 - 28)$ ☐ $(37 + 12)$	$(34 - 26)$ ☐ $(17 - 8)$

If ● + ● + ● + ● + 6 = 8 + ● + ● + ●, what is the value of ● ?

Test 1

Problem-solving

Harry Hiker saw 24 caterpillars when he was walking in the forest. When he stopped for lunch, he saw 5 more. How many caterpillars did he see altogether?

Wally Walker spotted and picked up 45 leaves on the grass. He then found 23 more in the park. What is the total number of leaves that Wally found?

Wendy Watcher saw 12 birds in her backyard today. Yesterday she saw 38 birds at the park. How many birds has she seen in all?

Tiffany owns 23 books. She received 2 more today for her birthday. What is the total number of books that Tiffany owns?

Tyler has 32 baseball cards. His friends give him 20 more. How many baseball cards does he have in all?

Christina counted 24 tulips in her garden. Her sister counted 15 roses. What is the sum number of flowers that they counted?

Anthony was on a Scouts trip and found 23 small twigs, 34 medium-sized twigs, and 25 large twigs for a campfire. How many twigs did he find altogether?

Tom, who was also on the camping trip, saw 10 squirrels climbing trees. He also saw 7 chipmunks scurrying to their homes. How many animals did Tom see in all?

Jennifer went fishing with her dad and she caught 3 fish. Her dad caught one. How many fish did Jennifer and her dad catch?

Larry, Curly, and Moe were playing a game of darts. Larry scored 10 points, Curly scored 12 points and Moe scored 8 points. What is the total number of points that they scored?

Yumiko helped her mother make four rolls of sushi. Her younger sister helped make two. How many rolls of sushi did Yumiko and her sister help make?

Dave found 17 chocolates in the community Easter egg hunt. He also helped his little sister find seven. How many chocolates did he find altogether?

Test 2

▲ − 7 = 16,　▲ + ● = 36

▲ =?

● = ?

▲ + ▲ = 16,　▲ − ● = 8

▲ =?

● = ?

▲ + ▲ = 24,　▲ − ● = 6

▲ =?

● = ?

6+ ▲ = 13,　▲ + ● = 11

▲ =?

● = ?

● − ▲ = 13, 15 − ▲ = 7

▲ =?

● = ?

If ● + ■ + ▲ = 21, ● + ■ = 13, ■ + ▲ =12

What values are ● , ■ , and ▲ each?

If ▲ + ▲ = 16, ▲ =?

▲ + ● =17, ● = ?

● + ■ = 19, ■ =?

Test 2

■ + ▲ =12, ■ + ▲ + ▲ + ▲ + ▲ =36

▲ =?

■ = ?

● − ▲ = 2

● + ● − ▲ = 10

● =? ▲ =?

▲ + ▲ + ▲ =33

● + ● + ● =21

▲ − ● =?

▲ − ● =10

▲ + ● =22

▲ = ?

● = ?

What are the even numbers which are less than 20?

Test 2

How many squares are in the following figure?

What are the odd numbers which are less than 20?

Ethan is in a lineup for a concert. There are 3 people in front of Ethan and 5 people behind Ethan. How many people are in the lineup?

Adam is 2 years older than Bob now. How old will Adam be more than Bob next year?

Meghan had 17 apples and she gave two of her friends every 4 apples then how many apples does she have now?

Andrew has $13 and he wishes to buy an applications software which costs $29. How much more he has to save in order to buy the software he wants to?

Test 3

Complete the pattern.

5, 51, 511, 5111, _____

Compute $17 + 15 + 13 + 11 + 9 + 7 - 16 - 14 - 12 - 10 - 8 - 6 =$

Complete the pattern.

10, 101, 1001, 10001, _____

Circle the number which does not belong to the group.

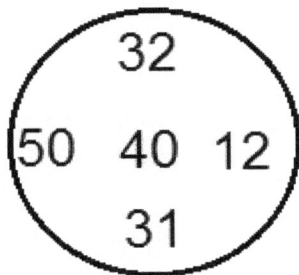

32
50　40　12
31

Find the next item.

Complete the next number.

88, 96, 11, _____

Test 3

Questions	Solutions
Tammy has 15 books. Tina has 36 books. How many books do they have in all?	
Mike has 37 stickers. Aaron has 19 stickers. How many fewer stickers does Aaron have than Mike?	
Sandy Liang has 29 e-mail messages. Jeffrey Lai has 13 e-mail messages less than Sandy. How many messages they have in all?	
Jun has 27 cookies and Veronica has 16 cookies more than Jun. How many cookies they have in all?	
Shirley has finished reading 119 pages of a book and there are 293 pages left. What is the total number of pages of the book in all?	
14 books were sold, and 29 books are left. How many books were there before selling?	

Test 3

Questions	Solutions
Amanda has 139 oil paintings and 2878 watercolour paintings, how many paintings does she have in all?	
The plumber has 219 copper pipes. There are 139 more copper pipes than steel pipes. How many steel pipes are there?	
Altogether, Michelle and Rosalind have 219 marbles. Michelle has 19 marbles more than Rosalind. How many marbles does Rosalind have?	
Rosalind has 321 marbles. Michelle has 51 marbles less than Rosalind. How many marbles does Michelle have?	
23 people got on the bus, 15 people got off the bus, 10 people got on the bus and 5 people got off and there were 23 people left on the bus. How many people were on the bus originally?	

Test 3

10 people got on the bus, 8 people got off the bus, 5 people got off and the bus was empty. How many people were on the bus originally (excluding the driver)?	
Jerry would have received as many e-mail messages as Julie if he had 16 more messages. Julie had 37 messages. How many messages did Jerry receive?	

Test 4

One chick has 2 legs.
Three chicks, how many legs?

Circle the odd one.

Which figure should be in place of the "?"?

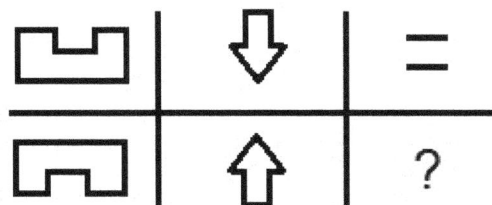

What number should replace ⓪ to balance the weights on the scales?

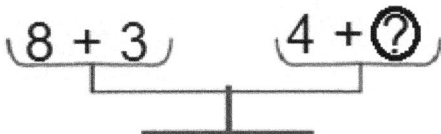

$$8 + 3 \qquad 4 + ⓪$$

I had 2 pairs of gloves, I lost one glove. How many gloves do I have now?

I am 8 years old and my brother is 5 years older than me. How old will my brother be next year?

How many rectangles can you see in the following figure?

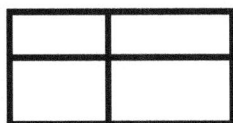

Student Name _____ Date _____

Test 4

After taking six plums from a basket, six were left in the basket. How many plums were in the basket at the beginning?

Bob has 8 oranges more than Adam and Adam has six oranges. How many oranges do they have altogether?

Amy has nine more apples than Bryan. Bryan has 3 fewer apples than Cathy. Cathy has 13 apples. How many apples does each one of them have?

Bill has four more apples than Bob and Bob has 8 more apples than Coco. Together, all three have 41 apples. How many apples does each one of them have?

Some birds were on the tree. Seventeen more birds flew back and eight flew away. Now there are 24 birds on the tree. How many birds were on the tree originally?

Fill in each the same box with the same number.

$6 + \square = 13$

$\square + \triangle = 12$

Test 5

If ☐ = 6 and ◯ =3, then ☐ + ◯ = ?.

5 = 7 – ☐

☐ + 3 – 1 = ☐ + ?

6 + 4 + ☐ = 13

7 + 7 + ☐ = 15

8 + 2 + 5 + 5 + ☐ = 21

A half dozen is 6. How many is one dozen?

Mom gave me one-half of $4. How much was it?

☐☐☐ – ☐☐☐☐ + ☐☐☐ = ? ☐

If ☐ = 2 and ◯ =3, then ☐ + ◯ = ?.

4 + ☐ = 6

One dog has 4 legs. If there are 12 legs, then how many dogs are there?

One boy has 2 legs. Two boys have _____ legs.

One wagon has 4 wheels, _____ wagons have 8 wheels.

If ☐ = 7 and ◯ = 5, then ☐ – ? = ◯ .

Test 5

If 13 + 7 + 6 + ⓐ = 31, then what is ⓑ ?

One boy has 2 legs and one dog has 4 legs. How many legs are there for 2 boys and 2 dogs?

The following large square is made of 16 small squares but some of them are missing. How many small squares are missing?

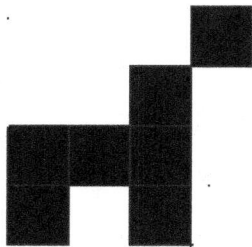

Each dot is two points. Find the value of one

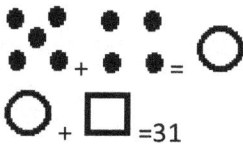

◯ + ☐ =31

What number should replace ⓐ to balance the weights on the scale 3?

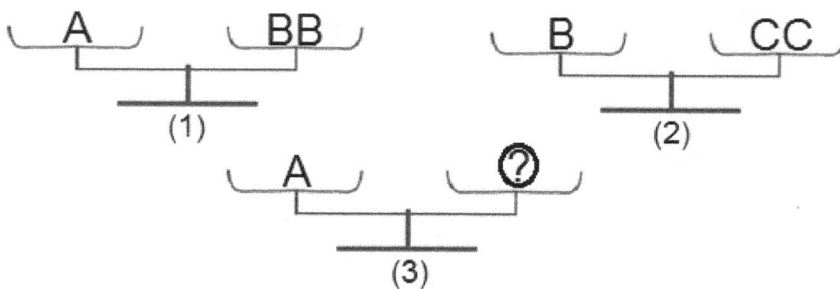

The first two scales are balanced. How many C's are needed to balance the scale 3?

Test 5

Which of the following figure has the longest line in the following figures?

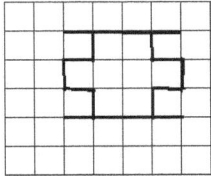

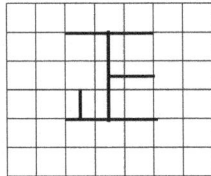

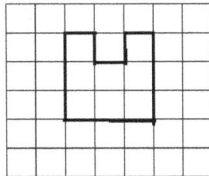

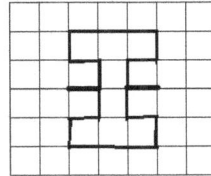

Last month, the cats Kiko and Snow together ate 12 cans of fish. This month they ate 5 more cans of fish than the last month. How many cans of fish did they eat in two months?

The cat Kiko likes to play pom pom balls but often she lost them and so far, she has lost 21 pom pom balls in two months. This month she lost twice as many as the last month. How many pom pom balls she lost last month?

The cat Snow likes to watch leaves falling from trees in the fall by turning his head up and down whenever he sees a leaf falling. This morning he saw 20 leaves falling and, in the afternoon, he saw one less than half of what he saw in the morning. How many leaves fell did Snow see in the morning and in the afternoon?

The cat Snow likes to jump up to catch his toy. In the morning, he jumped high 15 times which was one more than twice as many times as he jumped in the afternoon. How many times did he jump in the morning and in the afternoon altogether?

B

Find the values of the following figures.

○ + ○ + ○ + △ + △ + ▢ + ▢ + ▢ + ▢ =

423

○ = _____

△ = _____

▢ = _____

Fill in each square with a number from 1 to 9 such that the sum of each line is 9.

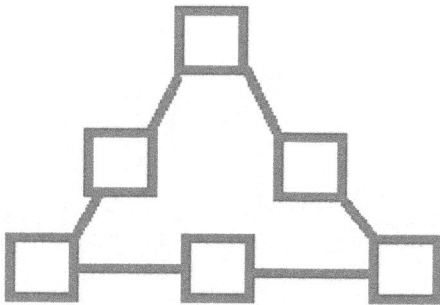

Part 12: A free Paper Set of Ho Math Chess

Want to have a free (almost) Ho Math Chess set? It takes about 10 minutes to complete a DIY chess set.

You can use the following link to print the chess pieces and the chessboard.
https://1drv.ms/u/s!AnVFh_KY48OgtJsze8k6ipl9GeaCQ

Chess pieces

Print the following Ho Math Chess set (on a colour printer is preferred), then cut each piece in a squared shape and paste them on pieces of corrugated papers, foams, or bottle caps.

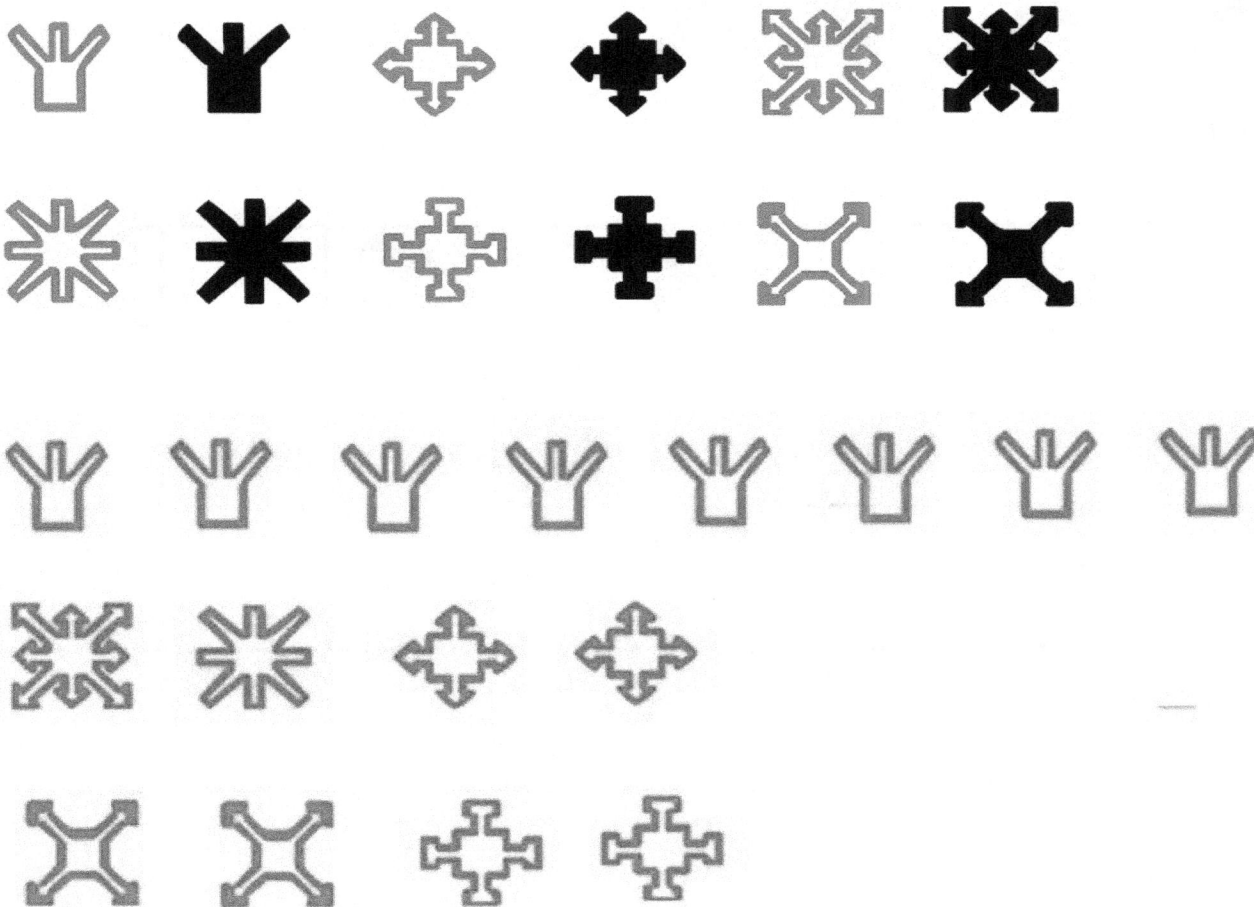

Chessboard

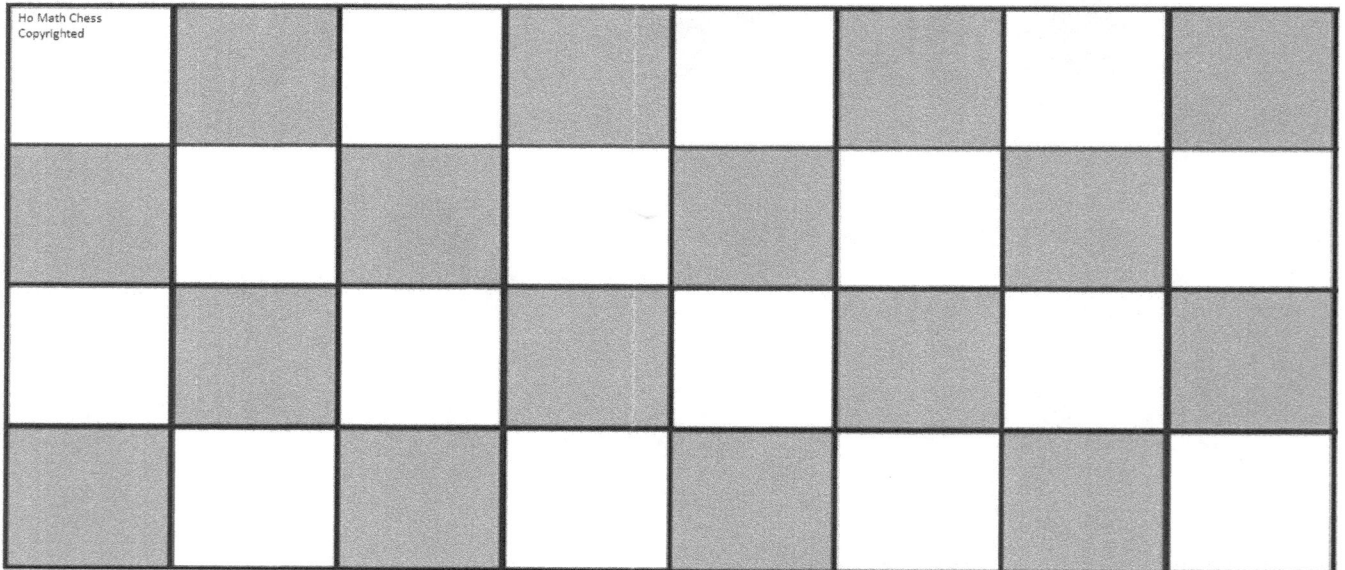

Ho Math Chess 何数棋谜　妈！我会棋谜式加法啦！

Mom! I Learn Addition Using Math-Chess-Puzzles Connection

Contents include both traditional and Math-Chess-Puzzles combined methods. Extra strength

©2008 – 2018 Frank Ho, Amanda Ho　All rights reserved. www.homathchess.com

Student Name _____ Date _____

A picture of DIY Ho Math Chess teaching set is imaged as follows:

介紹何数棋谜

何数棋谜 = 奧数棋谜 + 思唯腦力開發
英文教材，中英双语教学

什麼是何数棋谜?

上百篇科學論文巳發表國際象棋可以提高兒童問題解答能力. 並且訓練他們的專心及耐力. 所以我們巳經知道下國際象棋對兒童有好處. 但是因為國際象棋與計算能力並無直接開係, 所以如何讓兒童能在一個歡樂的環境下也能利用下棋來提高數學的計算呢? 何老師首創並發明有版权的幾何棋藝符號並利用此符號發明了世界第一的独特结合數學与棋谜教材. 何数棋谜讓兒童能利用幾何棋藝符號進行邏輯推理及數字的運算. 棋藝與算術的綜合題含蓋了整數, 幾何, 集合, 抽象數, 對比異同, 函數, 座標, 多空間圖形資料, 及規則性數字分析. 並且把棋藝的趣味性和數學的知識性結合在一起.

何数棋谜如何幫助兒童腦力思唯的開發?

很簡單的一個道理就是讓學生自願地去用腦, 何数棋谜首創獨一無二的融合數學與棋谜的独特趣味寓教於樂教材, 利用國際象棋訓練右腦的座標, 空間分析及圖形處理, 並利用發明了整合棋子與數學的圖形語言, 讓兒童能利用符號圖形訓練左腦進行邏輯推理及數字的運算. 國際象棋與算術的綜合題含蓋了整數, 幾何, 集合, 抽象數, 對比異同, 函數, 多空間圖形資料. 所以枯燥無味的計算題變成了謎題, 學生需要通過更多的思考. 能讓腦去思考愈多則腦力也愈開發. 處里訊息, 分析資料才能發掘出題目. 做這些謎題式數學時可以训練學生比較會專心及有耐心.

何数棋谜融合數學與國際象棋的教學理論巳在 BC 省數學教師刊物上發表. 科研報告已經證實何数棋谜教學法不但可以提高兒童數學解題及思維能力, 還可以開發兒童的腦力, 及分析問題的能力並且增加兒童學習的耐力, 學生的探索創造精神及求知欲. 判斷力, 及自信心等, 啟發思維訓練機警靈巧及加強手腦眼的靈活運用.

Introducing Ho Math Chess™

Ho Math Chess™ = math + puzzles + chess

Frank Ho, a Canadian math teacher, intrigued by the relationships between math and chess after teaching his son chess started **Ho Math Chess™** in 1995. His long-term devotion to research has led his son to become a FIDE chess master and Frank's publications of over 20 math workbooks. Today **Ho Math Chess™** is the world largest and the only franchised scholastic math, chess and puzzles specialty learning centre with worldwide locations. **Ho Math Chess™** is a leading research organization in the field of math, chess, and puzzles integrated teaching methodology.

There are hundreds of articles already published showing chess benefits children and that math puzzles are a very good way of improving brainpower. So, by integrating chess and mathematical chess puzzles together, the learning effect is more significant.

Parents send their children to **Ho Math Chess™** because they like **Ho Math Chess™** teaching philosophy – offering children problem-solving questions in a variety of formats. The questions could be pure chess, chess puzzles or mathematical chess puzzles in the nature of logic, pattern, tree structure, Venn diagram, probability and many more math concepts.

Ho Math Chess™ has developed a series of unique and high-quality math, chess, and puzzles integrated workbooks. **Ho Math Chess™** produced the world's first workbook **Learning Chess to Improve Math.** This workbook is not only for learning chess but also for enriching math ability. This sets **Ho Math Chess** apart from other math learning centres, chess club, or chess classes.

The teaching method at **Ho Math Chess™** is to use math, chess, and puzzles integrated workbooks to teach children fun math. The purposes of **Ho Math Chess™** teaching method and workbooks are to:

- Improve math marks.
- Develop problem-solving and critical thinking skills.
- Improve logic thinking ability.
- Boost brainpower.

Testimonials, sample worksheets, reports, and franchise information can be found at www.homathchess.com.

More information about **Ho Math Chess™** can also be found from the following publications:

1. Why Buy a **Ho Math Chess™** Learning Centre Franchise: A Unique Learning Centre?
2. **Ho Math Chess™** Sudoku Puzzles Sample Worksheets
3. Introduction to **Ho Math Chess™** and its Founder Frank Ho

The above publications can be purchased from www.amazon.com.